Cumbria Rocks

60 extraordinary rocky places that tell the story of the Cumbrian landscape

Ian Jackson

'.............to look at the scenery of Cumberland and Westmorland without trying to understand the rock is like listening to poetry in an unknown language – you hear the beauty of the sounds, but you miss the meaning. For the meaning is the rock.'

Norman Nicholson 1914 -1987

First published in the United Kingdom in 2022 by Northern Heritage Services Limited

Units 7&8 New Kennels, Blagdon Estate, Seaton Burn, Newcastle upon Tyne NE13 6DB
Telephone: 01670 789 940   www.northern-heritage.co.uk

See www.northern-heritage.co.uk for full catalogue

Edited by Angus Lunn, Abigail Burt and Tony Cousins

Graphic design: Ian Jackson and Abigail Burt

Mapping contains data from OS ©Crown copyright and database right (2022)
and ©OpenStreetMap contributors, Openstreetmap.org/copyright.

Printed and bound in the UK by W & G Baird

British Library Cataloguing in Publishing Data
A catalogue record for this book is available from the British Library.

ISBN 9781916237681

Cumbria Wildlife Trust

NORTHERN HERITAGE

WORLD LAND TRUST™
www.carbonbalancedpaper.com
CBP002879

FSC
www.fsc.org
MIX
Paper | Supporting responsible forestry
FSC® C016201

Fleetwith Pike
(Cover image: Grange Fell)

Eric Robson

"A splendid book that awakens millions of years of geological turmoil. The result is a thrilling tale that brings the events that shaped our landscapes to life."

Eric Robson OBE DL is a television broadcaster, author and documentary film maker who has lived for most of his life in Cumbria.

Foreword

Cumbria Wildlife Trust

As England's second largest county covering nearly 7000 square kilometres, a coastline of more than 180 miles – and boasting the 34 highest peaks in the country, Cumbria* has an unrivalled wealth of geological interest and an amazing variety of landscapes.

The wonderful array of rocks and landforms is not only important in its own right but provides the foundation for all of our habitats and wildlife. Geology has been central to Cumbria's history, economy, food and culture over millennia.

In this accessible and richly-illustrated book, Cumbrian Ian Jackson tells the fascinating and revealing stories of the county's geology through 60 special places.

Cumbria Wildlife Trust has always recognised the importance of geology and its conservation in our work. The promotion of a deeper understanding of geology has been a core objective for the Trust since 1962. This book, which celebrates our 60th anniversary in 2022, is the latest contribution to this mission.

Through our nature reserves but especially through the outstanding voluntary work of the Cumbria Geodiversity Conservation Group, we seek to record, conserve and promote the geology in our county. If this book sparks an interest, as I'm sure it will, please do join the Group and/or attend some of its events.

I hope this superb book will help everyone to appreciate the rocks beneath our feet and why they are so important to our past, present and future.

Stephen Trotter, Chief Executive, Cumbria Wildlife Trust

* Cumbria was created as a political entity in 1974 by the amalgamation of Cumberland and Westmorland and small parts of Lancashire. From 2023, two counties: Cumberland; and Westmorland and Furness, will be established.

Preface

Mention Cumbria and most people immediately conjure up an image of the Lake District. That's hardly surprising; its mountains and lakes are dramatic and magnificent. But almost two-thirds of Cumbria lies beyond the National Park and this too is a landscape of outstanding beauty and rich wildlife and heritage.

This book tells the story of the landscape of all Cumbria through 60 special places. There are many, many sites which could have been chosen. The places that made it into this book are a personal selection, ones which I felt would best explain and illustrate the evolution of the landscape of the whole county across geological time, and be accessible to visit. But I also wanted to emphasize the relevance of rocks to our everyday lives. That relevance can sometimes lead to controversy.

There are many excellent guide, photographic and geological books about Cumbria. Their focus has, understandably, been on the Lake District. It is also true that the explanation of the geology in guidebooks and photographic books is often minimal and sometimes wrong. Equally, there are many geological publications available and these contain detailed explanations which address the needs and interests of professional and amateur geologists.

The aspiration of this book, which follows one describing Northumberland, is to tell the story of Cumbria's rocks in a way that is understandable to everyone curious about the landscape of the county and to ensure that the information is scientifically correct and up-to-date.

I hope the book encourages you to explore all of Cumbria and, in Norman Nicholson's words, appreciate the meaning behind the beautiful scenery.

Ian Jackson

Vale of Eden

Sources and acknowledgements

The prime sources of information for this book are the maps, memoirs and reports produced by field geologists of the British Geological Survey (BGS).

I would particularly like to thank friends and former colleagues Brian Young, Tony Cooper, Dave Millward, Doug Holliday, Russell Lawley and Mike Howe for sharing their expertise.

BGS maps provide both comprehensive geological cover and essential detail and they are the foundation for virtually all scientific and applied research which seeks to understand the rocks of Cumbria and the UK. Many other books, scientific papers, articles, fieldtrip guides and web sites have also been consulted and I would like to acknowledge their authors, especially Clive Boulter for his excellent web synopsis of Lakeland Geology. May I pay tribute to the Geological Societies of Cumberland and Westmorland and the Cumbria GeoConservation Group for their continuing commitment to look after and make the geology of the county accessible.

Steve Trotter and Angus Lunn took on the challenge of providing information on the flora and fauna of several sites (within an unreasonable word limit!). Google, Wikipedia and many websites were searched to seek out information on the historical and cultural backstories. Reading a rich seam of natural history blogs was a real joy. Tony Cousins, Angus Lunn and Abigail Burt kindly proof-read and edited the text, but any errors in the descriptions are mine.

Thank you to my wife Gill and walking friends for allowing me to (again) hijack their hikes. The majority of the photographs were taken with my iPhone, but I'm hoping that Cumbria's stunning scenery will make up for that.

Finally thank you to Cumbria Wildlife Trust who generously supported publication of this book. All author's royalties will go to the Trust.

Lunedale

Introduction

Cumbria's rocks span almost 500 million years of Earth's history. From the time when life in primordial oceans was exploding, through periods of scorching red deserts and sub-zero Ice Ages right up to today and the sediments being deposited in our lakes, rivers and seas. All of these are part of our geological story*.

This diversity of the rocks makes Cumbria a unique and special place in Britain and is the reason that the landscape we see today has such amazing variety, from the fells of the Lake District, to the cliffs and marshes of the coast, the green pastures of our vales and the moorlands of the Pennine hills.

It is a diversity that has been exploited, by wildlife, by those mining for ore and coal but also by poets and artists like Wordsworth and Turner and today by millions who visit the county for its spectacular scenery and outdoor activities.

This book reveals the incredible story of the Cumbrian landscape through 60 special places and connects its rocks to wildlife, history, economy and culture.

To tell the story and make those connections the 60 sites have been divided into five themes but it is acknowledged that many could easily fit into several of these themes.

Some will argue, not unreasonably, that several sites in this book are hardly rocks. True, but collectively the 60 sites tell the full story of Cumbria's landscape and set in context the brief narrative of humankind.

Fleswick Bay, St. Bees Head

60 sites in 5 themes

Ancient rivers, seas & life
1 Christianbury Crags
2 Fleswick Bay
3 Gelt Gorge
4 Great Mell Fell
5 Holmepark Fell
6 Howgill Fells
7 Kentmere
8 Kingwater
9 Kirkby Stephen
10 Lacy's Caves
11 Lanercost Priory
12 Pardshaw Crag
13 Parton

Volcanoes & molten rock
14 Armathwaite
15 Carrock Fell
16 Ennerdale
17 Eycott Hill
18 High Cup Gill
19 High Rigg
20 Pavey Ark
21 Shap
22 Threlkeld
23 Tilberthwaite

Earthquakes & folded rocks
24 Arnside
25 Brothers Water
26 Causey Pike
27 Dufton Pike
28 Hartside
29 Mosedale
30 Penton Linns
31 Shap Summit
32 Tarn Hows
33 Trusmadoor

Climate & landscape change
34 Brampton
35 Glasson Moss
36 Glenridding
37 Great Asby Scar
38 Hale Moss
39 Helvellyn
40 Inner Solway
41 Mickleden
42 Ravenglass
43 Skelsmergh
44 Slater Bridge
45 Wastwater

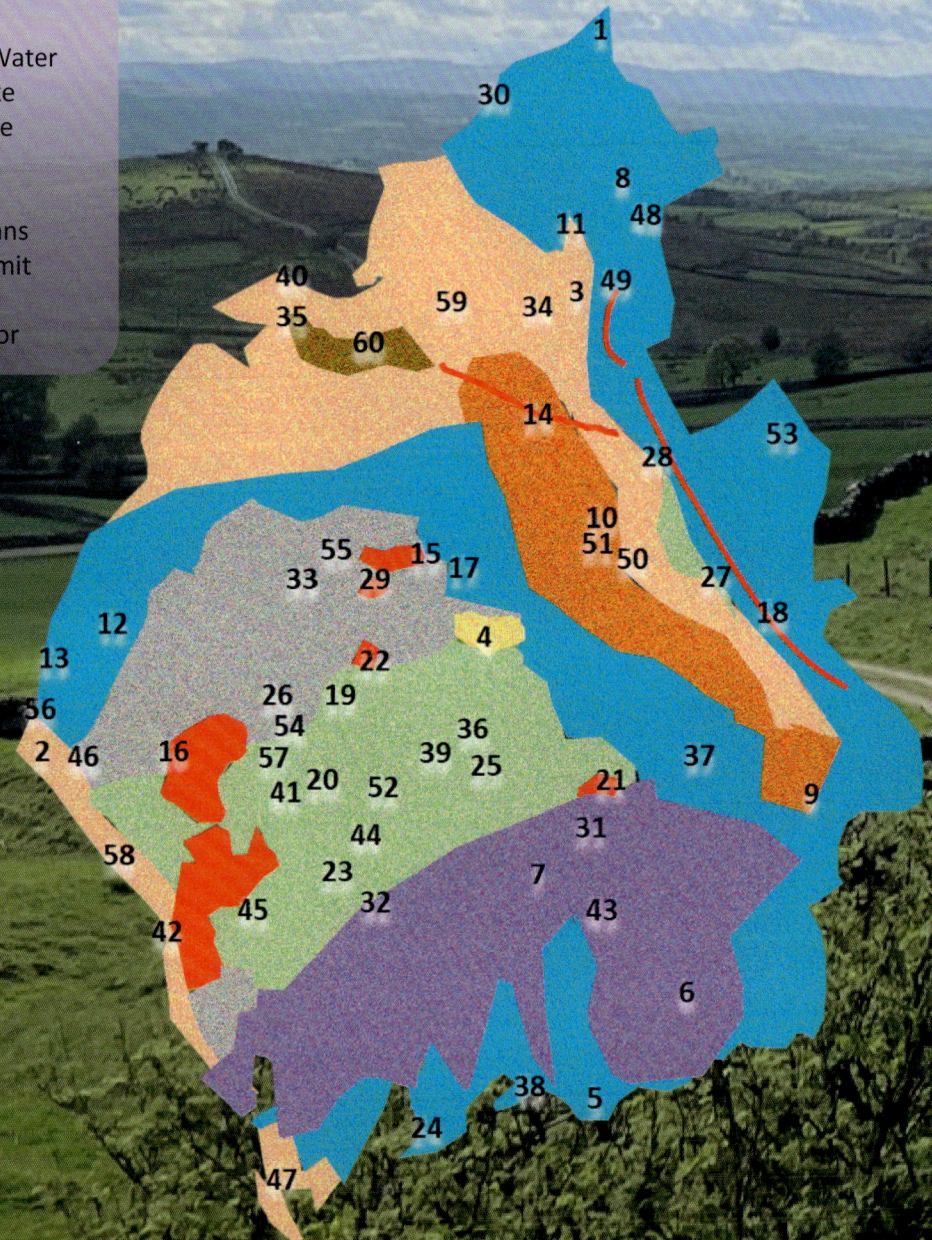

Heritage & mining
46 Florence Mine
47 Furness Abbey
48 Gilsland Spa
49 Hallbankgate
50 Long Meg
51 Long Meg Mine
52 Loughrigg
53 Nenthead
54 Newlands
55 Roughton Gill
56 Saltom Pit
57 Seathwaite
58 Sellafield
59 Tullie House
60 Watchtree

A66 near Brough

Cumbria's geological journey

Cumbria and its rocks started their journey close to the South Pole around 485 million years ago. Back then Cumbria was between two continents and at the bottom of a deep ocean which was filling with mud and sand. The tectonic plates carrying the continents grew closer, forcing one deep beneath the other

The ocean continued to fill with sediments and eventually closed, while forces within the Earth buckled the crust into a huge mountain chain

Around 350 million years ago our itinerant piece of the tectonic plate was almost at the Equator and our environment had changed too: large rivers ran through tropical swamps and into warm coral seas. Sea level rose and fell as the Earth's ice sheets grew and then receded

2.6 million years ago the Earth's temperature began to fluctuate again, we cooled and then warmed, repeatedly. Ice sheets extended to lower latitudes, glaciers grew and then retreated and as they did trees and plants re-colonised the landscape. Humans emerged and began to exert their influence on the county, clearing forests and settling in the valleys. But it is society's development and our impact on the planet in the last 150 years that far outstrips the previous 300,000 years of our existence

Pressure and heat melted the rocks and the molten magma not only filled huge chambers deep underground, which were to become granite, but erupted vast amounts of volcanic ash and stones

Millions of years of erosion wore those mountains down and covered a now subdued desert landscape in sand and gravel

Our drift north continued and desert conditions returned. Earth movements created tension in our northern rocks and deep down hot molten magma was injected along cracks and fractures. On the surface sand dunes migrated across a hot plain and flood-prone rivers deposited even more red sand and silt

The tectonic plates continued their erratic waltz and 60 million years ago lava started to pour out of an enormous rift that was to become the Atlantic Ocean. North America and Europe have been drifting apart ever since

For the next 150 million years a sea teeming with life covered the county completely. But of that sea and its rocks there is little or no sign, for other than a small remnant, all that too was eroded away

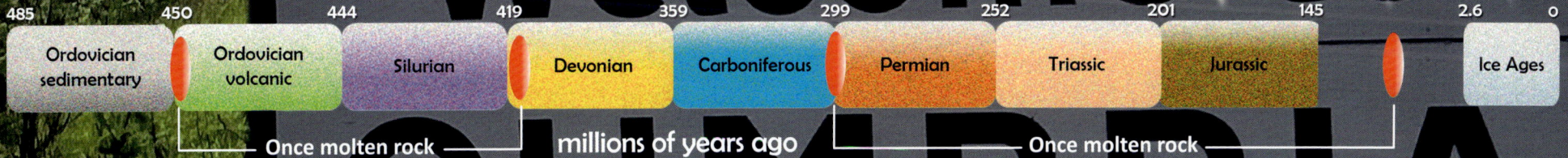

485	450	444	419	359	299	252	201	145	2.6	0
Ordovician sedimentary	Ordovician volcanic	Silurian	Devonian	Carboniferous	Permian	Triassic	Jurassic		Ice Ages	

— Once molten rock —   millions of years ago   — Once molten rock —

Christianbury Crag

In the remote far north corner of Cumbria, surrounded by miles of Sitka spruce trees, stands a lofty crag. The views into Scotland and across Cumbria are expansive.

The landscape looks the same as the extensive sandstone uplands of Northumberland which lie only a few hundred metres to the east. As you might expect, Christianbury Crag shares the same geological history as its Northumberland cousins. It is also Carboniferous and around 340 million years old. Back then, as the grains and sloping layers in the rocks show, they were sand in a fast-flowing river draining to a southern sea. Burial under thousands more metres of sediment and uplift into mountains followed and only after millions of years of erosion did the crags we see now appear. Their disjointedness and rugged profile today owe much to the attrition and melting of ice sheets only 20,000 years ago and their pot-holed surface to weathering since. The crags are a small example of a 'rock city' with sandstone blocks separated by corridors.

Before the mass planting of conifers began here in the 1920's (a post-war move to secure a timber supply for the nation), Christianbury Crag was a prominent landmark. The 'Famous Christenbury Craigs' featured in a 1754 edition of 'The Gentleman's Magazine' with a full-page engraving and was compared in stature to the manmade edifice of Stonehenge. It was described in Victorian travelogues as 'rising grimly from the heathery waste, a haunt of foxes'. Walter Scott's fictional characters frequented these rocks, as did real life murderers and their gamekeeper victim, Thomas Davidson, strangled in 1849. Today the crag is defended by thick forest and set in a superb blanket bog dominated by sphagnum mosses, heathers, cotton grass and deergrass. There are ring ousel, wheatear and whinchat around the rocks and breeding red grouse, dunlin and golden plover on the moor. Reaching the crags is a challenging uphill hike along forestry tracks and boggy footpaths. If you're seriously fit you might also want to take in Glendhu Hill, a candidate for England's most remote hill and pay your respects at Davidson's monument on the way back.

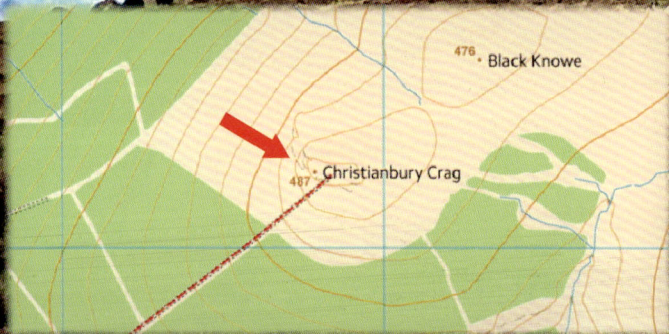

There's a car park near Cuddyshall Bridge and it is a 16 kilometre rough roundtrip hike east to the Crag [NY578823]

View northwest over the eroded sandstone escarpment of Christianbury Crag

Fleswick Bay

Fleswick Bay, St Bees Head, is one of the best and most accessible places to see the St Bees Sandstone and to understand the way it was formed.

For these reasons St Bees Head is designated as a Site of Special Scientific Interest (an SSSI). The red sandstone in the sheer cliffs, and forming the rock platform of the foreshore, is magnificent. The rock is from the Triassic Period, around 250 million years ago. The way the size of the sand grains changes within the rock layers and the angles of the layers tell a story of sand and mud being carried down braided rivers flowing across a desert plain. Sometimes the rivers flooded violently, at other times the flow was much less. The rocks show ripple marks and cracks where mud has dried out. When these rocks were being formed England was only about 25 degrees north of the Equator; the same latitude as the deserts of Arabia today.

The wave cut rock platform has been eroded into strange but beautiful undulating patterns. They look like ripples but are most likely caused by the intersection of natural cracks in the sandstones, which result in lines of weakness. Above the rock platform, and moving every day with the tides, is a beach of small, rounded pebbles made of many different rock types. These stones were brought here by ice sheets that flowed down the Irish Sea and across northern England 20,000 years ago, then they were tumbled and polished by the sea.

This area is an SSSI for its wildlife too. The cliffs are home to guillemots along with fulmars, kittiwakes, razorbills, cormorants, puffins, shag, herring gulls, tawny owls, sparrowhawks, peregrines, ravens and rock pipits. There is sheep's bit in the coastal turf; it's very much a western British plant.

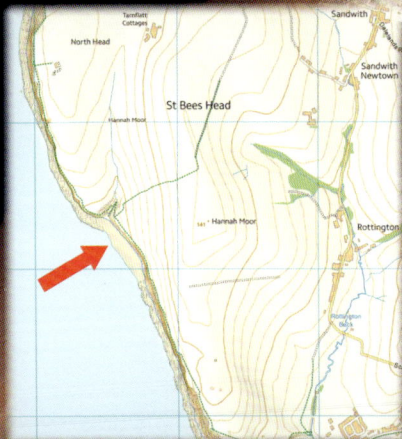

There is a train service to St Bees, or you could park near the beach. It's a 3 kilometre walk from there [NX945133]

The Triassic sandstone cliffs and wave-cut platform of Fleswick Bay

Gelt Gorge

The River Gelt flows down from the northern Pennines and then, to the northwest of Castle Carrock, descends into a beautiful red sandstone gorge. The rocks in the gorge were once sands in a desert, but that was 250 million years ago.

These sandstones, red but sometimes white and green, are from the Triassic period of Earth's history. Geologists call them the St Bees Sandstone formation because they occur there too. The sandstones were mostly deposited by water flowing into temporary lakes in the desert, but some of the rocks were deposited by wind as well.

There are at least two Roman quarries in the sides of the gorge. The stone is easily shaped and carved and, as well as for building Hadrian's Wall, may well have been used for monuments and architectural purposes. The soldiers who worked the quarries left many wonderful carvings that are visible from the path – letters, names and even caricatures of their officers.

The area has a gruesome history. In the 16th century there was a bloody battle near Hell Beck Bridge between the armies of Elizabeth 1st and Leonard Dacre, a local rebel lord. Nearby, at the Capon Tree, six of Bonnie Prince Charlie's men were hanged after the 1745 rebellion. That turmoil is in the past; today the gorge is a beautiful place to walk. The River Gelt has cut winding channels, chutes and potholes in the sandstone and Gelt Woods is one of few surviving ancient woodlands in East Cumbria. Even though the canopy trees are much-altered by human activities, the shade and humidity provide ideal conditions for mosses, ferns, fungi and a varied ground-flora.

Walk up the gorge from the car park at Low Gelt Bridge [NY520592]

Roman carvings in a quarry in St Bees Sandstone in the Gelt gorge

Great Mell Fell

Once Britain had rocks called the New Red Sandstone and the Old Red Sandstone; Great Mell Fell is a bit of the old sort.

Because they looked the same (red and mostly sandstones), the Old Red and the New Red used to confuse geologists. Nowadays these rocks and their very different ages are better understood and we call the 'New' Permian-Triassic and the 'Old', Devonian, named after rocks in Devon. We can't be sure about their exact age because there are no fossils in Cumbria's Devonian rocks, but they are thought to be from the younger part, around 380 million years ago, when mountain torrents carried more than 1500 metres thickness of cobbles and sand onto a desert plain. The alluvial fans and braided, gravel-filled channels that these flash floods created have now become a rock called conglomerate. That's what Great Mell Fell and its neighbour Little Mell Fell are made of.

If you hike up Great Mell Fell from the southeast, or hunt around in the stream that flows through the woods, you will find bits of conglomerate, as well as many of their cobbles and stones made of older Lake District rocks. Above the woods, there are larch and pine trees; their bent boughs are testimony to the exposed position of the fell and the strength and persistence of the prevailing westerly wind. But that exposure means the view of the mountains of the Lake District from the summit, and especially of Blencathra, is worth the climb.

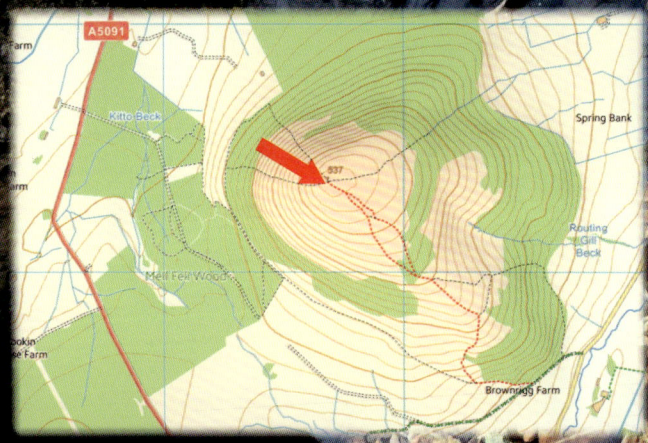

Just north of Brownrigg Farm there is a small pull-in. The walk to the summit is about 2 kilometres [NY397254]

View southeast over Great Mell Fell

Holmepark Fell

Just after you cross into Cumbria heading north on the M6, you'll maybe notice a prominent pale grey hill to the east. It's called Holmepark Fell and you might also see it's got a notch across its top.

The pale hill is limestone but the notch is (was) a rock layer called the Woodbine Shale. Climb up the hill and you won't find the shale; it's hidden by limestone scree. Shale and its close relative, mudstone, are the invisible men of field geology. You know they exist; the quarry next door has excavated them; boreholes prove they are there; and you sometimes catch a glimpse in a fresh cliff. But in the landscape shales rarely reveal themselves. Reason: shales are easily eroded. They were once mud and while they may have been compacted, they are the soft option. Limestone and sandstone are much harder, they resist erosion and stand proud in the landscape. Shales give us the depressions and valleys, the negative features. Yet shales make up a lot of the rocks in Cumbria, especially those from the Carboniferous Period. At Holmepark Fell the Woodbine Shale has a thickness of around seven metres but the notch you can see is where it used to be. Time and weather have worn it away. But the view from Farleton Knott is well worth the walk.

Shales are in the public eye, controversially. Natural gas can be found trapped in tiny pores in shale. The gas comes from the decomposition of ancient plants and animals and can contain 70 to 90% methane. The technique used to extract shale gas is called fracking; drilling boreholes and artificially fracturing the rock to release the gas. The Bowland Shale, which stretches underneath much of Lancashire and Yorkshire, is a Carboniferous shale like the Woodbine Shale, but it is one hundred times thicker and is rich in organic material. It is a potentially valuable prospect for gas production and that means in these times of rising energy costs a contentious debate might re-start.

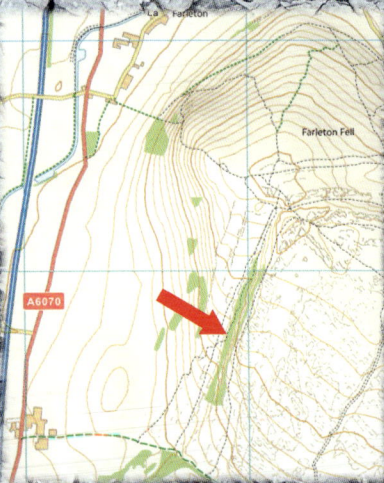

Take the bridle path from Holme Park Farm onto access land of the fell. It's a very pleasant 2 kilometre return walk [SD539797]

The limestone escarpment of Holmepark Fell; the Woodbine Shale lies hidden by scree

Howgill Fells

Travelling north by rail or road, perhaps the first view you get of Cumbrian fells are the Howgills just south of Tebay. They look steep and lonely and they are.

These big green whalebacks seem to have little rock, few walls and fences and fewer people, so they are quite different to their Lakeland cousins. But that's their appeal, plus they have different aspects of Cumbria's geological story to tell. The bedrock, usually hidden beneath a veneer of glacial clay and debris, is mostly from the Silurian Period. Thousands of metres of sandstones, mudstones and siltstones are evidence of the infilling and closure of the ancient Iapetus Ocean as two continents collided and one dived below the other 415 million years ago. The sandstones (called turbidites) often have grains of very different sizes and this is because vast amounts of debris slid quickly off the continental shelf into a deep ocean. These catastrophic submarine slides still happen today and occasionally they cut through cables carrying the world wide web. While the Howgill Fells look peaceful enough now, their geology continues to evolve. Clay and stones left by the ice sheets 20,000 years ago, also creep easily down the steep slopes and increased rainfall, helped by people and their animals changing the vegetation, can produce rapid erosion and landslides. If you are lucky you may catch sight of wild fell ponies in the lonely valleys.

Right on the southern margin of the Howgills, where a popular path to Cautley Spout waterfall begins, is the Cross Keys Inn. A place with Quaker associations, it has been a temperance inn since 1902. There is a William Blake quote carved on a lintel 'Great things are done when men and mountains meet. They are not done by jostling in the street.' Few geologists would disagree.

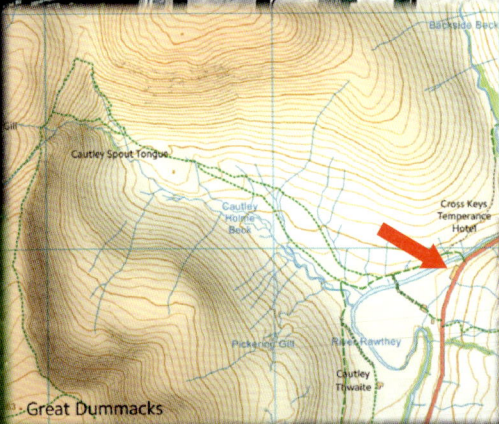

There is a pull-in east of the Inn and it is a 2 kilometre walk to Cautley Spout from there [SD681975]

View southwest over Great Dummacks and Cautley Spout

Kentmere

What are smaller than the width of a human hair, can double their population every 24 hours, generate about 35% of the world's oxygen and used to live in Kentmere?

The answer is diatoms, a form of single-celled algae that live mostly in water, including the oceans, and absorb silica to build their skeletons. They thrive in cold, clear water and colonised the original Kentmere lake (which gave the valley its name) around 10,000 years ago after the last glacier left. While diatoms are found in many upland lakes in Britain the huge volume and purity of the diatoms in Kentmere was unusual. This is probably down to the unique position and shallowness of the lake, the lack of sediment that flowed into it and perhaps the abundant silica in the volcanic rocks to the north. Several metres thickness of diatoms collected at the bottom of the lake and their accumulated skeletons became geology: diatomaceous earth. Sometime in the mid-19th century Kentmere was drained to try to create more farmland. It wasn't a success and the old lake bottom remained boggy. In the 1920's two geologists accidently discovered the diatomaceous earth. Its physical properties made it valuable; it is inert, very light, and doesn't conduct heat well. Excavation probably began in the 1930's and, as the current tarn progressively formed, work continued by dredging. The diatom skeletons were dried, crushed and then used in many things, from filters to abrasives to insulation. Working stopped about 1980, and at its maximum 10,000 tons was being extracted each year. The men who worked there also discovered two dug-out canoes in the old lake bed; one found in 1959 is probably Viking and you can see it in Kendal Museum.

Diatoms are not just geological or biological curiosities. Our modern demand for miniaturisation of components means that scientists are presently researching how diatoms reliably and repeatedly manufacture their complex skeletons to create amazingly resilient nano-scale structures.

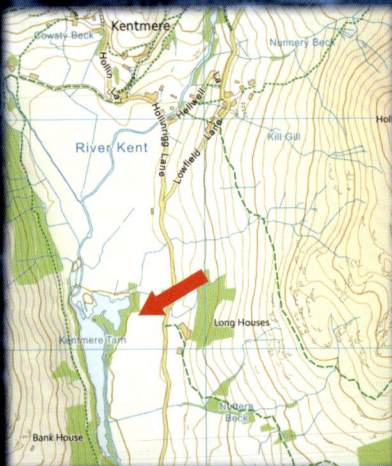

Parking is in short supply in the valley, but you get a view of the Tarn via the footpath to its west [NY455026]

Micrograph of Kentmere diatomaceous earth. Permit number CP030/22 British Geological Survey © UKRI 2022. All rights reserved

View north over Kentmere Tarn

Kingwater

North of Gilsland is a small stream; Kingwater. Sitting in the bed of the stream, exactly where they grew 340 million years ago, are nine fossil tree stumps: the best example of a grove of Carboniferous trees in life position in Britain.

The largest stump of these primitive trees is almost 2.5 metres in diameter, so they grew to a considerable size. They are called Pitus primaeva and when looked at under the microscope show a cell structure typical of a type of seed ferns called Lyginopterid pteridosperms. The organic material in these stumps has, over time, been replaced by dolomite and calcite, minerals which percolated down from limestones above and preserved the anatomy of the trees. As you might expect, fossils like these, which can tell us much about the environment so long ago and how early plants and trees developed on Earth, are very rare and very delicate. This is a Site of Special Scientific Interest and if you visit, please take only photographs.

Despite, or perhaps because, there is an RAF Electronic Warfare facility close by, the upper reaches of Kingwater are quiet and secluded, sheltered by commercial conifer plantations. Roe deer are abundant, there are wild brown trout and if you are very very lucky you may see a goshawk.

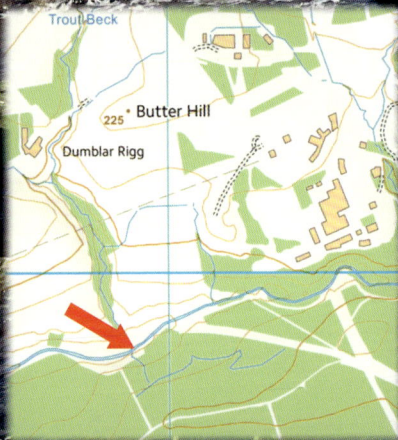

Image of thin section of fossil tree under microscope

Take the road to Spadeadam and park before the no entry signs. Kingwater is just over a 1 kilometre walk northwest from there [NY609698]

Stump of fossil tree in the bed of Kingwater

Kirkby Stephen

In the valley of the River Eden beside Kirkby Stephen is a rock called Brockram. That's a local Cumbrian word and means broken rock, which is exactly how it looks.

Angular fragments, mostly of grey limestone, but with some sandstone and rare Whin Sill dolerite, are bound together in a pale red, sandy, muddy cement. The geological term for rocks like this is breccia. The fragments are mostly older Carboniferous rocks and sometimes you can find characteristic fossils of corals and other marine animals in them. Around 275 million years ago in the Permian Period these fragments were scree and debris eroded from a nearby mountain range and carried quickly downstream by flash floods onto a plain below. At that time Britain was around the same latitude that Oman is now, and the environment and its climate was the same too: hot, dry, mountainous desert with abrupt rainstorms.

Beneath Stenkrith Bridge the River Eden has cut a spectacular narrow slot channel (called the Span of Eden) through the Brockram. You can also see Brockram in old quarries west of Nateby, probably the source of some of the stone used to construct many of Kirkby Stephen's distinctive buildings.

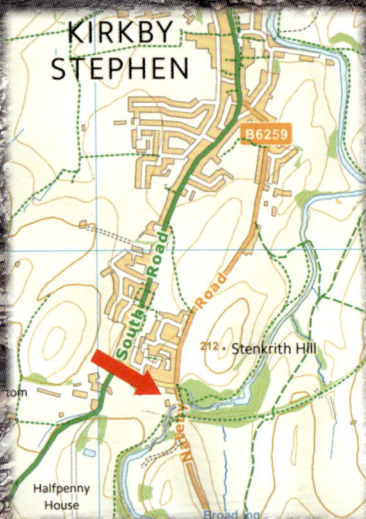

There are rail and bus services to Kirkby Stephen. From there it is a short walk along riverside footpaths [NY773074]

The 'Span of Eden' eroded in the Brockram beneath Stenkrith Bridge

Lacy's Caves

In the Vale of Eden, from just south of Carlisle, in a band passing through Penrith and Appleby to Kirkby Stephen, a red sandstone dominates both natural rock outcrops and the building stones of towns, villages and farms.

The rock is called the Penrith Sandstone and it is Permian in age, that's around 275 million years ago. Lacy's Caves are just one of many places you can see it. If you look at the sandstone with a magnifying lens you can see many rounded grains of quartz, with a red, iron-rich, coating. The surface of the grains is frosted (finely pitted), telling us that this was once sand blowing around in a desert. In other outcrops, such as Cowraik Quarry, you can even see the dunes, formed by a prevailing wind that blew from the southeast. Desert environments at that time hosted few animals and rarely preserved fossils, but footprints of vertebrates (perhaps amphibians or reptiles) have been found in slabs of sandstone nearby.

At Lacy's Caves, like many other places, the sandstone is soft enough to be carved, but elsewhere it can be a hard and durable building stone. The difference is that the hard rock has silica holding the grains together; a secondary solution process that happened not long after the dunes were formed. Diagonal white veins that crisscross the rock were also once a silica solution.

Who was Lacy? He was a Lieutenant Colonel who lived in Salkeld Hall in the 18th Century. Like others at the time, he was attracted by the notion of romantic nature and so had the caves carved; it's even claimed he installed a hermit there to complete the folly. His empathy with landscape didn't extend to Long Meg stone circle however; he is said to have planned to have it blown up but if he did that failed.

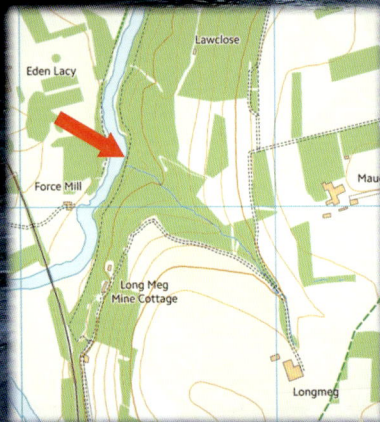

Close-up of rounded and frosted sand grains

Park in Little Salkeld and take the footpath north beside the railway line and the river. It is a 5 kilometre roundtrip walk [NY563380]

Silica veins criss-crossing Penrith Sandstone in Lacy's Caves

Lanercost Priory

If you can't see bedrock, the stone used in local buildings is a great clue to the geology. Lanercost Priory goes one better, it tells us where the join is between two major geological periods of Earth's history.

The Priory is a 50:50 blend of two sandstones: a younger, dull red Triassic (St Bees) sandstone and an older pale grey-brown Carboniferous sandstone. The colours of these two different rocks add unique character to the stunning buildings. They also tell us about different past environments. The red sandstone was deposited in temporary rivers and lakes in a desert 250 million years ago and the grey-brown sandstone was once sand in rivers that ran through tropical swamps 330 million years ago. You can't see it in the landscape here but a geological map shows that the boundary between the two rocks runs almost due north-south through Lanercost.

The Priory was founded in 1169 by Augustinian canons. They used stone from local quarries to the west and to the east. But, along with many other church, castle and house builders, they also robbed ('re-purposed') stone from Hadrian's Wall, another monument whose building stone changed as it crossed the Triassic-Carboniferous boundary. The Romans quarried stone at many sites along the Wall and two of them, in the gorge of the River Gelt and at Combcrag, upstream in the Irthing valley, are less than six kilometres away. There are Roman inscriptions in both quarries and on the re-used stones in the Priory. Lanercost has a rich and sometimes turbulent history. It was plundered by William Wallace and Robert the Bruce and became a resting place for Edward I, 'The Hammer of the Scots', before his death. It is an evocative place.

The Priory is an English Heritage site and you can park beside it [NY555637]

Triassic and Carboniferous sandstone building stones in Lanercost Priory

Pardshaw Crag

To the southwest of Cockermouth is a small village called Pardshaw. It lies on the side of a limestone ridge, just south of a craggy hill. The crags offer a splendid view of the north-western fells.

335 million years ago, in the Carboniferous Period, this limestone was a mix of limey organic mud, shells and corals in a tropical sea. England was close to the Equator then, not 55 degrees north like now. Over aeons our tectonic plate has steadily moved across the globe. During that time the mud and shells were deeply buried and compressed into rock. Millions of years of Earth movements and erosion have now revealed them. While Carboniferous rocks like these today appear like a ring around the Lake District, millions of years ago they, and the younger rocks above them, completely covered all of the high fells.

Below Pardshaw Crag sits an historic meeting house of the Quakers; it is the place where John Dalton, the Eaglesfield man credited with developing atomic theory, went to school in the 1770's. But around 1650, years before the present meeting house was built, the founder of the Quakers, George Fox, preached to thousands of Friends from Pardshaw Crag. When you visit you can see what a wonderful natural pulpit it is. Limestone bedrock produces a very particular set of plants, including thyme and common rock-rose and grasses like sheep's fescue.

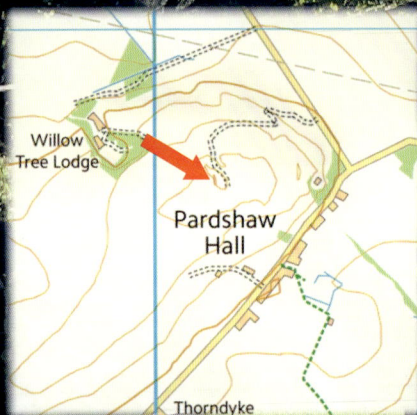

Willow Tree Lodge

Pardshaw Hall

Thorndyke

The crag is 300 metres northwest of the Quaker Meeting House [NY103256]

View south towards the Lake District mountains from Pardshaw Crag

Parton

Parton and its shoreline have two histories. One is recent and the other is ancient, but they are closely connected.

Around 315 million years ago, this part of Cumbria was a hot tropical swampland with large rivers that often overflowed their banks, pouring sand and mud into lagoons. The muds gradually settled often covering the leaves and stems of plants that grew beside the rivers. Many of the plants in the swamp became peat and as more sediments built up over millions of years they were compressed into coal. The sands were compacted into resistant sandstone and form the cliffs and foreshore from Saltom Bay northwards. The mud is now shale and it preserves those leaves of primitive ferns and giant club mosses as fossils. You can search for them, and other Carboniferous plant and animal remains, in stones on the beach.

There are many other rocks on the beach but not all of them are natural. The coal that was once a hot peaty swamp became the foundation of the West Cumberland economy from the 18th to the 20th century. Coal mines, and the iron and steel manufacturing they sustained, have long gone. What remains are glimpses of their existence in the industrial archaeology along the coast and its debris on the beaches. Parton was once a thriving coal port, with a tannery and glass-making factory. It was a rival to Whitehaven until a storm in 1795 destroyed its harbour. By the end of the 19th century the town was in decline and many houses became slums. They were eventually demolished and Parton moved up the cliff and away from the sea. Workington-born artist Percy Kelly painted Parton. Kelly was a friend of poet, Norman Nicholson and like him was someone whose art captured perspectives from the industrial rim of 20th century Cumbria – a very different approach to those focused on the scenic and picturesque in central Lakeland.

There is a railway station in Parton (a request stop) and a car park beside the beach [NX979206]

Extract from a painting of Parton by Percy Kelly, by kind permission © Chris Wadsworth

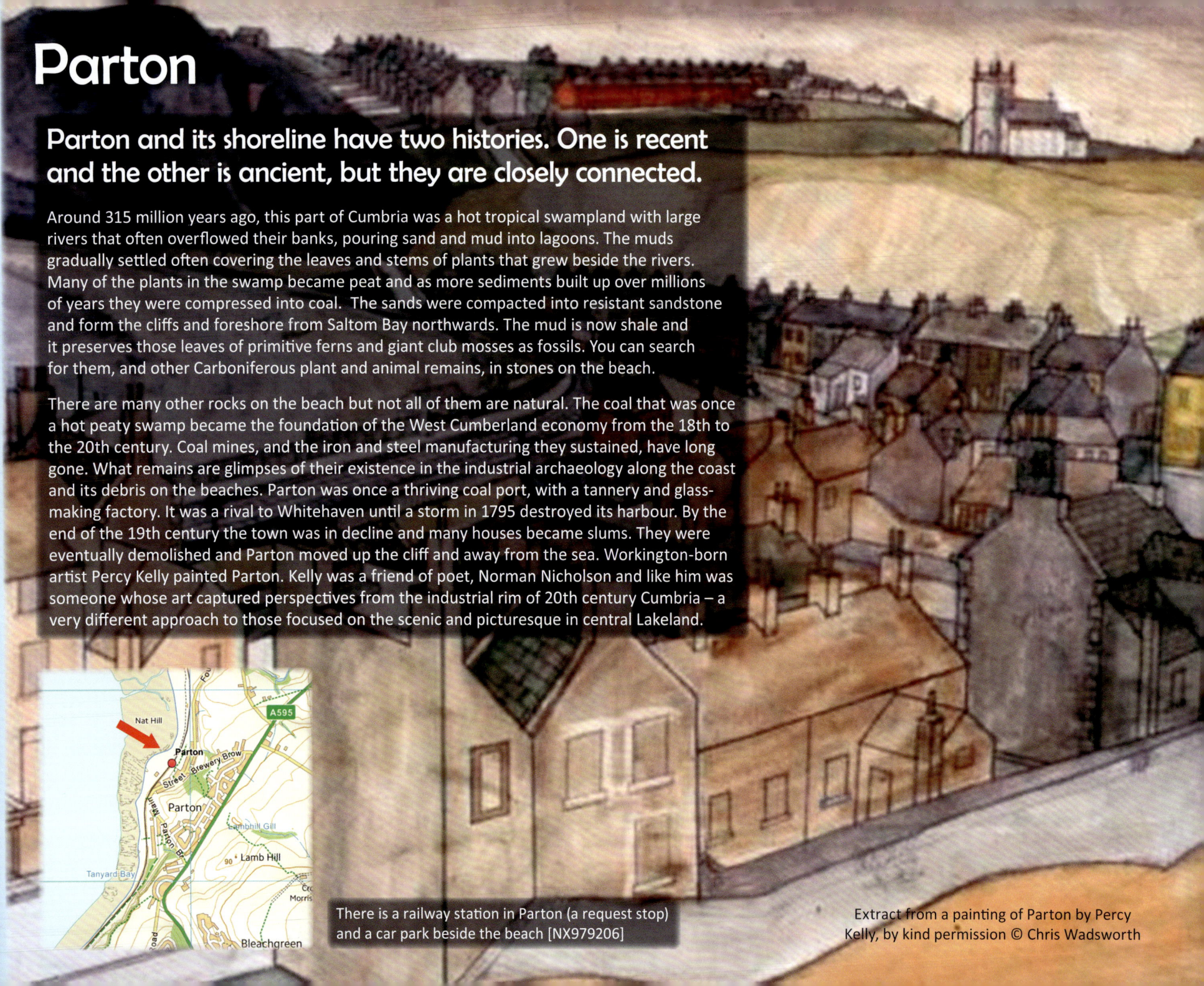

Plant fossils in Carboniferous mudstone

Armathwaite

The youngest bit of solid rock in this book stretches northwest - southeast for many tens of kilometres across the whole county but it is never more than 35 metres wide. It is called the Cleveland-Armathwaite Dyke.

The best place to get up close to it is where the dyke cuts across the River Eden south of Armathwaite; here the relative hardness of the rock, in contrast to the softer red sandstones either side of it, creates a natural weir. That hardness also means you can also see it in the landscape. It is the foundation of the chain of long, low hills you see between the M6 and the Pennine fells as you drive between Penrith and Carlisle. The rock is igneous, meaning it was once molten; it is a fine-grained 'andesite' intruded into much older rocks around 60 million years ago. It is one of several near-vertical blades of rock that are thought to originate from beneath the Isle of Mull in western Scotland. The theory is that a single pulse of molten magma spread laterally at a velocity of up to 18 kilometres per hour so injection of the whole dyke would have taken less than five days.

A belt of wildlife-rich native woodland hugs the east bank of the River Eden on the edge of Coombe Wood. This ancient wood was planted-up with conifer trees in the 1950s but today a presumption in favour of native broadleaved species is benefitting wildlife.

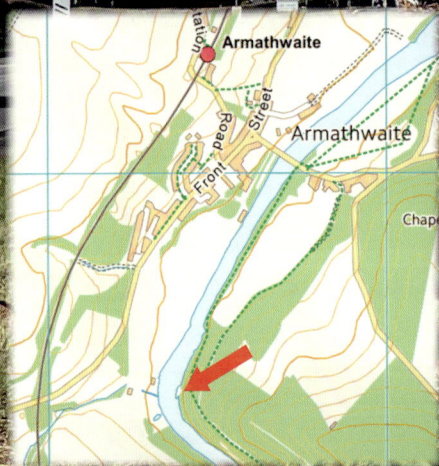

Park near Armathwaite bridge and take the footpath south along the east bank of the river for 800 metres [NY504454]

The dyke creating a natural weir south of Armathwaite

Carrock Fell

Right on the north-eastern edge of the Lake District is a fell whose profile is recognisable from miles away. It's special geologically too.

Carrock Fell is made up of a lot of different rocks and one of them, gabbro, is rare in Cumbria, in Britain and on the surface of the Earth. Most igneous rocks (rocks that were once molten magma) that form the Earth's continental crust contain lots of silica making them relatively light in weight and in colour. But gabbro has lots of iron and magnesium and is darker and heavier. Gabbro and its volcanic equivalent, basalt, are common in the oceanic part of the Earth's crust. They are also present where continents break apart and have origins in the Earth's deeper layers. At Carrock Fell and in the more famous Black Cuillins in Skye, gabbro began as molten rock, several kilometres beneath the Earth's surface. Around 460 million years ago it was injected (intruded) as a series of layers into older volcanic rocks. Directly to the north of the gabbro and cutting through it are younger silica-rich rocks: microgranites, which are lighter in colour. Both these sets of rocks were later broken up by stresses in the Earth. That fragmentation and their complicated chemistry, has made their story difficult to unravel and is why, for a long-time, the area was simply referred to as the Carrock Fell complex. It continues to fascinate geologists today.

Carrock Fell has other stories to tell. In Grainsgill, in the valley below, are the remains of one of only two tungsten mines in Britain, a legacy of the special chemistry of these rocks and the adjacent Skiddaw granite. On the top of the Fell is a rampart of stones, an Iron Age hill fort; it has a wonderful view but it must have been a cold stronghold.

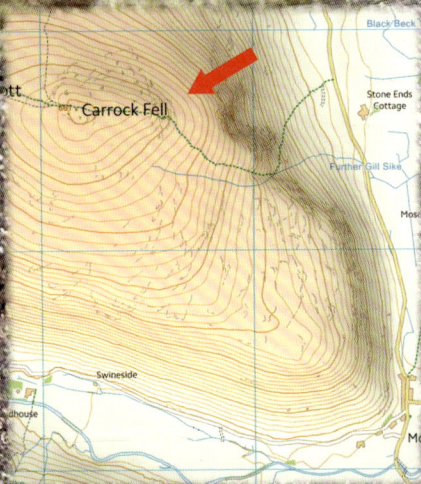

There is parking at Apronful of Stones immediately east of the Fell. It's a short, but strenuous, hike to the top [NY342336]

Carrock Fell viewed from the southeast

Ennerdale

If you were to ask someone who knows the Lakes well in which valleys would you find most peace, they would probably include Ennerdale.

Cumbria's most westerly finger lake sits in a beautiful glacial valley. Its location and the absence of public roads are the reason for its tranquillity. At the valley's western end are 475-million-year-old, bent and fractured mudstones and sandstones; go east of Angler's Crag and the bedrock is Ennerdale Granite. Once called the Ennerdale Granophyre, detailed work by a British Geological Survey team in the 1980's and 90's, covering all Lake District rocks, produced a more up-to-date identification. Technically the rock is a 'granophyric microgranite'; which means it has small, interlocking mineral crystals. Geophysical and geochemical investigations showed the microgranite to be around 1-2 kilometres thick, to extend over 53 square kilometres and to have been intruded around 450 million years ago. At the head of Ennerdale the conspicuous mounds are moraines, deposited by a small glacier which reformed in a short return of a cold climate about 12,000 years ago.

Ennerdale's secrets extend to much more than its rocks. The lake is home to the rare Arctic Char, a throwback to Ice Age times and a fish more common in lakes in the north of Canada. On the bed of the River Ehen, the valley's outflow, are equally rare Freshwater Pearl Mussels. Their rarity and environmental sensitivity were the catalyst for a recent £300 million alternative water supply scheme to ensure the lake would continue to sustain the river. The Wild Ennerdale project has ambitious plans to re-wild the valley, including re-introducing beavers. On the floor of the valley near Low Gillerthwaite are mounds of stones. These are a 'cairnfield' or clearance cairns, evidence of the creation of pastureland by the clearing of forest by Bronze and Iron Age peoples 5000 and 3000 years ago. The cairnfield here is now threatened by erosion by the River Liza; as ever geological processes are indifferent to human accomplishments.

Parking is at the west end of the lake. For a stunning view climb Angler's Crag [NY099150]

View of Ennerdale from above Angler's Crag

Eycott Hill

This is a Cumbria Wildlife Trust nature reserve. It is also a Site of Special Scientific Interest because its rocks as well as its wildlife are of national importance.

So Eycott Hill is a name well-known to geologists. The rocks here, layers of lava flows, are a key chapter in the geological history of Cumbria. They erupted out of fissures and vents during the Ordovician, around 450 million years ago, as a fiery precursor to the closure of the ancient Iapetus Ocean and the impending collision of continents. The lavas are mostly a fine-grained grey rock called andesite, but some of the rocks contain holes and pink blobs; bubbles of gas in the lava which can be filled with silica. There are even older rocks on Eycott Hill; 480-million-year-old buckled mudstones that were once mud in that ancient ocean. Geologists have found primitive microscopic fossils called acritarchs, which are more common in the earlier Cambrian Period of Earth's history. That would make these the oldest rocks in Cumbria! There are younger rocks in the reserve too, 150 million years younger. Limestones from the Carboniferous lie between the hill and the road to the east. Sink holes caused by acid rainwater dissolving the limestone are the clue.

Eycott Hill's rocks (its geodiversity) have produced the varied habitats we see today. The very irregular topography, with small ridges and intervening wet depressions, is called 'knock-and-lochan' (after similar features in the Scottish Highlands). It was formed by ice sheet erosion emphasising the differences in rock strength. There are areas of rough upland pasture, rocky dry knolls and peat bogs and fens between the knolls. That variety means this is a place rich in animal and plant life. Notable are uncommon sedge species like bog-sedge and few-flowered sedge.

Park in the Trust car park and walk 500 metres west to the Reserve [NY387295]

Eycott Hill lava with vesicles formed by gas bubbles

18

High Cup Gill

Due northeast of Appleby is a huge straight gash in the Pennine escarpment, 2.5 kilometres long, almost a kilometre wide and 200 metres deep.

It's called High Cup Gill but is often known as High Cup Nick. The Gill is a glacial valley, cut during successive ice ages. Around the upper rim is a 30-metre cliff of rock called the Whin Sill (miners named it: whin because it's hard and sill because it's horizontal). The Whin Sill is made of an igneous rock called dolerite and was injected into the older Carboniferous rocks as molten magma around 295 million years ago. It is estimated that the Sill originally had a temperature of 1100°C and took 60 years to cool. The rock contracted as it cooled, forming those characteristic vertical cracks and joints.

The Whin Sill (or Sills as there are several parts to the intrusion) extends over a huge area of northern England, from Bamburgh on the Northumberland coast, to Hadrian's Wall and the moors of Teesdale. With an area of at least 4500 square kilometres, most of it beneath the surface, and an unknown area beneath the North Sea, it is England's biggest intrusion and the original sill of geological science.

Molten rock and glacial erosion are not the only things of geological interest in High Cup Gill. Towards its western end the 350-million-year-old Carboniferous rocks lie directly on 475-million-year-old Ordovician rocks. That 125-million-year gap represents millions of years of erosion and is something geologists call an unconformity; multi-million-year gaps like this are a common feature of rock sequences everywhere. On damp crag faces can be found fir clubmoss and, indicating the basic nature of Whin minerals, mossy saxifrage and hoary whitlow-grass.

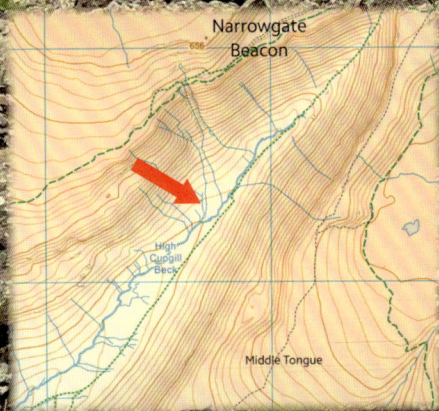

Narrowgate Beacon

High Cup Beck

Middle Tongue

There is parking in Dufton village. The Nick is a strenuous 12 kilometre return hike [NY745262]

View west down High Cup Gill

19

High Rigg

In Cumbria between 450 and 460 million years ago volcanoes spewed ash and lava over a vast area and red-hot clouds of molten debris flowed down slopes at hurricane speeds.

Most of the central Lake District is made of these ancient volcanic rocks but it always disappoints to have to explain that, despite their triangular shape, mountains like Bow Fell and Catstye Cam are not themselves volcanoes but are huge chunks of volcanic terrain that have been broken apart, re-arranged and then eroded. So the eastern face of High Rigg is a geology teacher's dream; in clear view as you drive through St John's in the Vale, is one of the very few places where that primeval volcanic terrain is obvious at landscape scale. The sloping crags and benches are alternate layers of harder lava and more easily eroded volcanic breccias (a rock made of angular fragments). The lavas, an igneous rock called andesite, are from some of the earliest eruptions in the Lake District. They are part of a unit called the Birker Fell Formation which spreads over 315 square kilometres. The landscape then was more like Hawaii than Krakatoa; these are shield volcanoes, vast sheets of molten magma bursting from fissures and flowing down low-angled slopes.

High Rigg's wildlife is almost a microcosm of the wider Lake District fells. Its slopes are covered by coarse bent and mat grasses with just a few heathery shrubs and flowering herbs such as the white heath bedstraw or the yellow tormentil. Bracken is common and may reveal where the broadleaved 'rainforest' woodland once grew. High Rigg is one of the 'Wainwright' fells and an easy one to climb from St John's church; the views south are worth it.

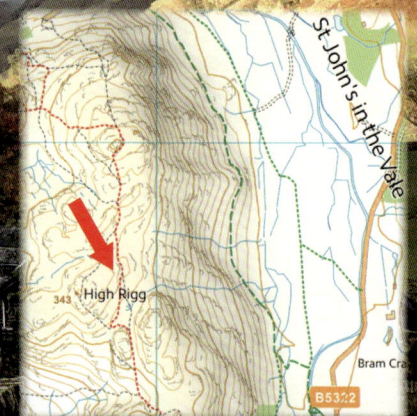

St John's Church has a small car park and the summit of High Rigg is a 1 kilometre walk south from there [NY307214]

View southwest over High Rigg fell

20

Pavey Ark

The rocks of the Scafell range and the Langdale Pikes record the most violent periods in the volcanic history of Cumbria.

Geologists estimate that over 400 cubic kilometres erupted from a huge volcano before its crater (called a caldera) collapsed and filled with water and more volcanic debris. The fantastic rocks on Pavey Ark are the product of this highly explosive eruption. Red hot debris, ash and gas surged down the slopes to the lake at great speeds. These are called pyroclastic flows and form rocks called ignimbrites and volcanic breccias. The weird and bent shapes in this photograph are spatters of lava, once hot and soft enough to be deformed. The breccias contain lava blocks full of holes that once contained gas. There are also blocks ripped off the walls of the volcano. The crater lake progressively filled with this mix of debris all of which had a volcanic origin. They are called volcano-sedimentary, or volcaniclastic rocks because, as their layered and laminated textures show, these are also sediments deposited in water.

An enormous range of rock types were generated by this ancient volcano 455 million years ago. Then these rocks were dislocated, broken and weathered over millions of years and it is this combination of processes that has given us some of the most rugged and stunning landscapes in the Lake District.

Pavey Ark is a serious summit and a strenuous 6 kilometre return hike. Ascend Stickle Ghyll and approach from east [NY285079]

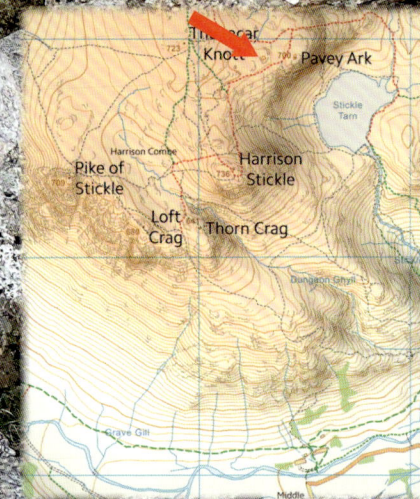

Explosive volcanic rocks above Langdale Valley

21

Shap

What's the best-known rock in Cumbria? Shap Granite!

It's pretty distinctive with large pink crystals and you don't have to visit the quarry to see it because it's been used as a decorative building stone across the country. The rock is made of different minerals: feldspar (pink and white), quartz and biotite. Around 405 million years ago, in the Devonian Period, it was molten magma about eight kilometres underground. It cooled slowly giving time for those large crystals to grow. Since then, millions of years uplift and erosion have brought the granite to the surface. In the last 2.6 million years ice sheets have picked up blocks and boulders of Shap Granite and moved them across the country. Because the rock is so easy to recognise, these itinerant blocks (erratics) have helped us understand the directions that the ice took. Shap Granite boulders have been found down the Vale of Eden, in the valley of the River Tyne, in the Yorkshire Dales, and even on the east coast at Bridlington, 160 kilometres away.

You can't go far without seeing Shap Granite as a building stone in our northern towns; Metro stations in Newcastle city centre are faced with it. It was so popular that you will also see it in St Mary's Cathedral in Edinburgh, at the entrance to St Pancras Station, on the Albert Memorial and even in the Precinct of St Paul's Cathedral. Shap granite often has areas of a different, dark-coloured rock in it, bits from the older surrounding rock that got caught up in the molten magma. Geologists call them xenoliths; quarrymen call them heathens. I wonder if there are any in St Mary's and St Paul's?

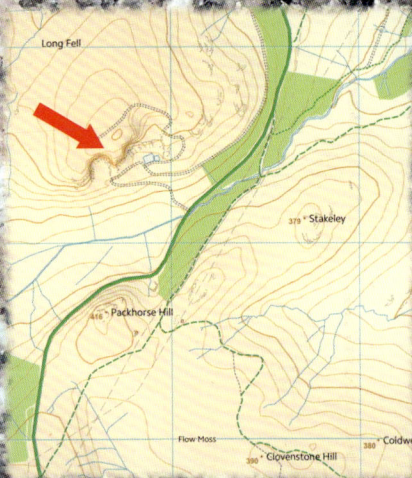

Shap Granite quarry is a working site. You can see it from the A6 or hike across the granite on the fells around it [NY559085]

Erratic boulder of Shap Granite beside Wet Sleddale Reservoir

22

Threlkeld

The village of Threlkeld lies in the shadow of Blencathra. To the south of it, at Threlkeld Knotts, there is a granite named after it. How it got to be there is an enigma.

Geologists call Threlkeld granite a microgranite because its mineral crystals are small. It was once molten magma intruded into older mudstones and volcanic rocks and it's part of a larger, 1500 square kilometre, granite body (called a batholith) found deep beneath the Lake District. Using the length of time radioactive uranium takes to decay into lead, scientists have dated the granite at around 450 million years old making it Ordovician. There is more to this story. Geologists believed that the shapes of Threlkeld Knotts, and Clough Head behind it, could be explained by intrusion of the magma, faulting and erosion, especially by glaciers. But recently other Earth scientists have argued that a catastrophic landslide; a huge 'rock slope failure' related to the last ice sheet 20,000 years ago, is the reason why it looks like it does today. Looking south from the A66 you can appreciate why the big landslide concept is tempting, the dramatic scar on the face of Clough Head is plain to see.

Quarrymen in the late 19th century didn't debate such niceties, they recognised the microgranite as a rock that was hard but wasn't the best as a polished stone. So they crushed it and used it for rail ballast and construction instead. The quarry closed in 1982 and it is now the home of Threlkeld Quarry and Mining Museum. Visit and learn about the equipment and the people who worked there, and about a time when mining and quarrying put bread on the table for many communities in Cumbria.

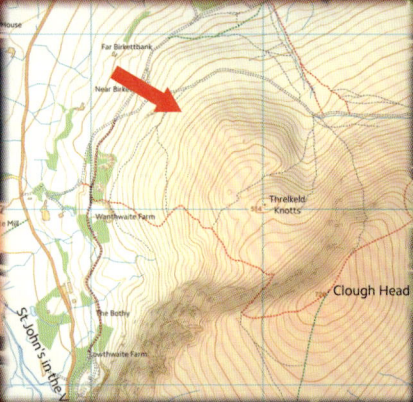

Visit the Threlkeld Quarry and Mining Museum, or climb Threlkeld Knott [NY329230]

View of Threlkeld Knotts and Clough Head from the north

23

Tilberthwaite

There used to be hundreds of mines and quarries in the Lake District. The little valley of Tilberthwaite was home to copper mines and slate quarries but only one small one is still working slate today.

The beautiful green slates you see on Lake District roofs, but more often nowadays as table mats and name plates, are rocks with a complicated story. The fact that they are included in a set of rocks known as the Borrowdale Volcanic Group is one complication. Quite often only part of their origin is volcanic; the other part is sedimentary. They are called volcano-sedimentary and they are a hybrid made up of hot dust, ash and rock fragments ejected from a volcano and then transported by water, wind or gravity. Those journeys result in some beautiful patterns in the stone, ripples that tell of flowing water currents, and even shaking by earthquakes.

There is another complication with slates; the flat surface of a roofing slate is not the original horizontal "bedding" surface when this was mud and volcanic debris around 450 million years ago. The rock suffered huge pressures within the Earth for millions of years afterwards and in response the particles were flattened at a completely different angle. It is this "pressured" surface, called cleavage, along which the slate naturally splits.

The vast 18th and 19th century roofing slate industry has gone and green slate is now the basis for a more select business: high value home and garden products which grace shop windows in Ambleside, Keswick and Grasmere. But you don't have to buy to see green slate, just look down; some of the most beautiful examples which tell the stories of their origin make up the paths you walk along and the walls beside you.

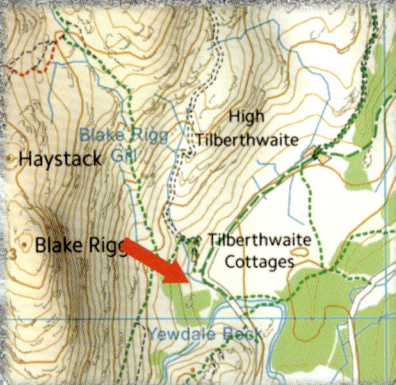

Lots of parking beside the old quarries [NY306010]

Disused slate quarry at Tilberthwaite

24

Arnside

Just inside the southwest border of Cumbria is Arnside, a peninsula with some great folded and broken rocks.

Arnside is part of an Area of Outstanding Natural Beauty (Arnside and Silverdale AONB). It overlooks Morecambe Bay and coastal cliffs two kilometres west of the village provide the chance to get close up to some splendid geology: evidence of ancient Earth movements that deformed these rocks and there are fossils too. The limestones in these rocks were once limey mud together with the shells and skeletons of animals that lived in a warm coral sea at a time when Britain was located just south of the Equator. This was around 340 million years ago during the Carboniferous. You can see the preserved remains of some of those animals (like the coral here but also shellfish called gastropods and brachiopods) in rocks that have been polished by the sea.

Because the sea cliffs cut through and expose the layers of limestone it makes it easier to appreciate the forces at work. North of Park Point are places where you can see both faulting and folding of the rocks. The shallow curved shape of the coast between Park Point and Silverdale is caused by a major fault-line. Most of the breaking and bending we can see took place around 290 million years ago during the 'Variscan Orogeny', a period of mountain building on a continental scale. But these lines of weakness were inherited from a much older time of upheaval in the Earth's crust; 100 million years before.

Arnside's rich wildlife extends from the wading birds that thrive on the sands and muds of the foreshore and estuary, to inland habitats often underpinned by its limestone foundations, with an array of woodland birds, butterflies and wildflowers.

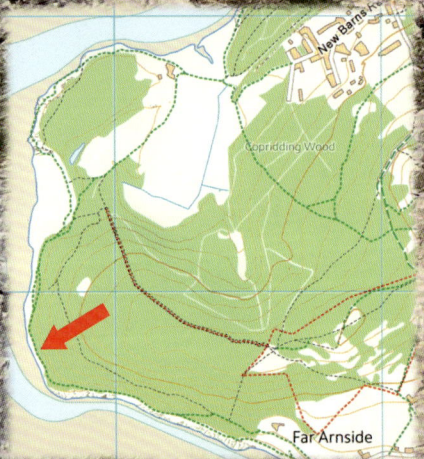

Fossil of coral in limestone polished by sea

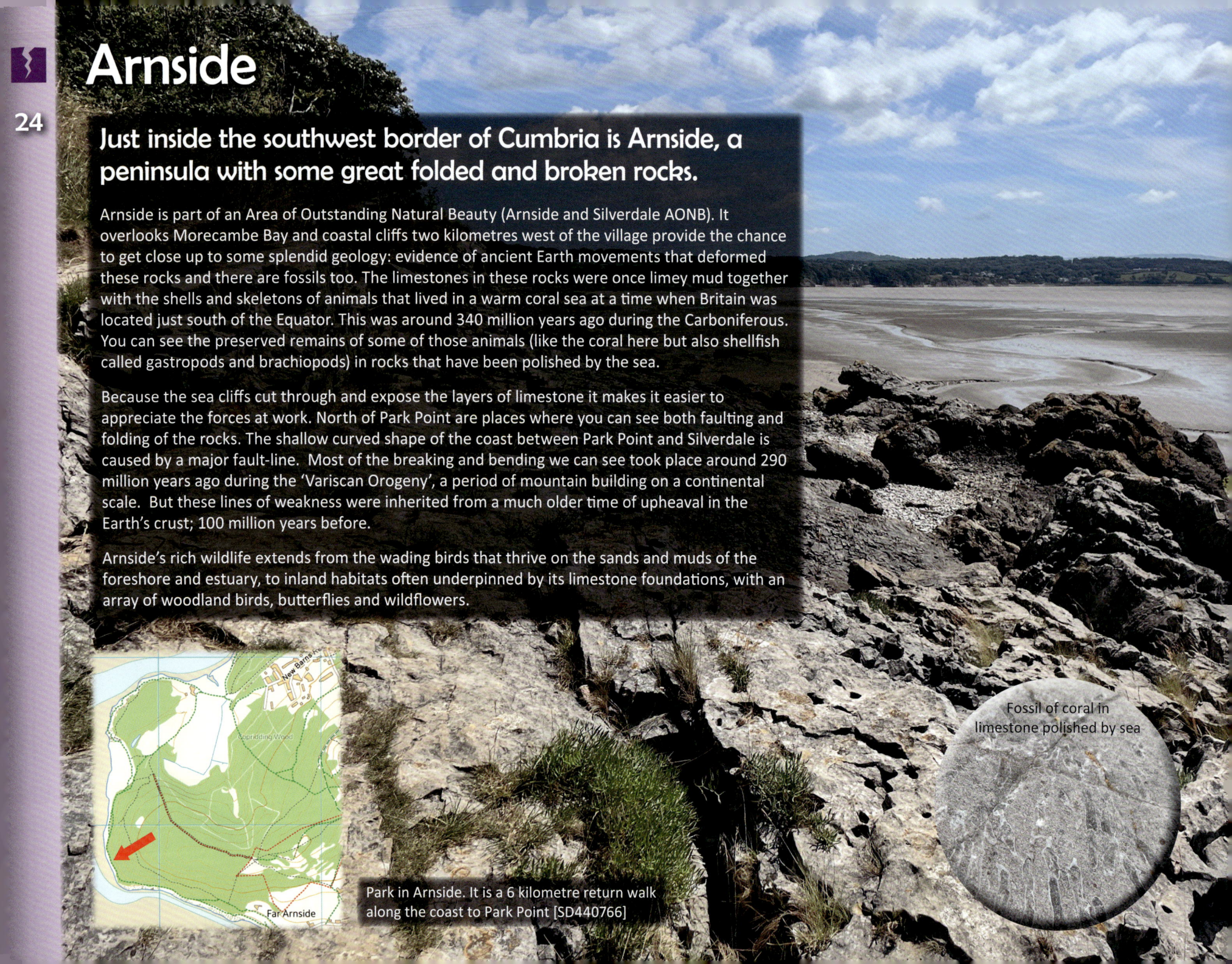

Park in Arnside. It is a 6 kilometre return walk along the coast to Park Point [SD440766]

Geological fault in limestone cliffs at Arnside

25

Brothers Water

Gaze north down Kirkstone Pass. The valley and its steep, ruler-straight sides are striking, especially the west shore of Brothers Water.

Why is this valley, and many others in the Lake District, so linear? The explanation is geological. Events many millions of years apart have produced the landscape we see today. Several episodes of Earth movements caused faulting which has broken these 450-million-year-old rocks into a jigsaw of fragmented pieces. The fault lines became zones of weakness and rivers that flowed long before the Ice Age, eroded their courses along them. In the last 2.6 million years successive glaciers followed those river valleys and carved deeper and broader 'U' shaped troughs. When the glaciers melted, they disgorged vast amounts of sediment which produced the flat valley bottoms; something that the rivers and streams continue to do today. Brothers Water and the valley's straight lines are because of large faults on either side and because of the action of rivers and glaciers that followed them. In a final act debris flowing down Hayeswater Gill through Hartsop created a barrier of sand and gravel across the valley floor. The story of how this landscape evolved is based on many decades of geological research. But like all geological explanations, it is an interpretation of data available today: data about the shape of the land, about rock types, and our knowledge of how processes within the Earth and on its surface work. As we gather new data our understanding and explanations may change, that is the way science progresses.

The name Brothers Water has a tragic origin. It used to be called Broad Water, but two brothers were said to have drowned in the late 18th century when they fell through the lake's frozen surface. They were possibly not the first brothers to do that. You could see goosander, cormorant, common sandpiper and coot on the lake and there is a typical northern meadow on the eastern shore with globeflower, wood crane's-bill and great burnet.

You can park at Cow Bridge and walk south or look down on Brothers Water from Kirkstone Pass [NY402131]

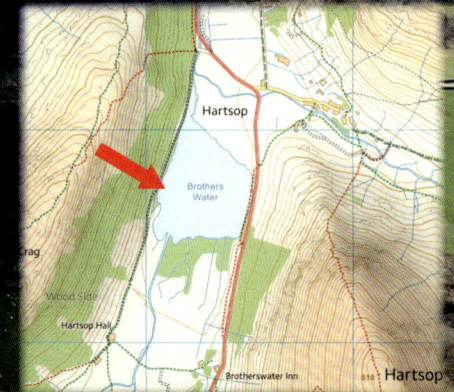

View of Brothers Water and the faults that define the valley from Brock Crags

26

Causey Pike

It was probably Alfred Wainwright who said you get the best views from little fells. That works for geology too.

Climb Barrow or Ard Crags and look over at Causey Pike. Filling the view from east to west is a formidable ridge. From Causey Pike's distinctive knoll it runs over Scar Crags, Sail, Wandope and Whiteless Pike. Just south of the crest of the ridge is a major break in the Earth's crust, a geological fault, but not a normal one. It is a thrust fault; one where older rocks to the north have been pushed up over younger ones to the south. While different in age the rocks either side of the fault belong to the same geological period, the Ordovician. They are Skiddaw Group mudstones and sandstones and have suffered quite a bit of pressure and heat over their 480 million years. The Causey Pike Fault is huge, it extends more than 45 kilometres from Ennerdale in the west to Cross Fell in the east, and dislocates rocks by as much as two kilometres. For a closer look, take the footpath up Causey Pike from Uzzicar and walk along it!

Just like the Skiddaw Group rocks it divides, this fault and others just to the north, have a complex story. The faults may have originated when huge volumes of sediment being deposited along the margins of the ancient Iapetus Ocean slumped to the deeper sea bed. 70 to 80 million years later a major episode of mountain building re-activated weaker slide-zones in the slumps and moved them again. After enduring such cataclysmic events perhaps it's not surprising that these rocks look so bent, broken and cleaved.

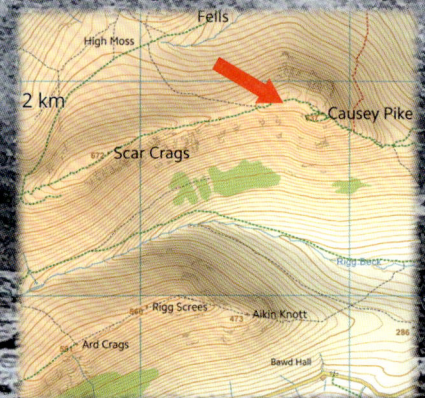

There is off-road parking at Stoneycroft. It is a 5 kilometre strenuous hike to the summit [NY219209]

The line of the thrust fault just south of the summit of Causey Pike

27

Dufton Pike

From the A66 near Appleby they look like three green pyramids lined up in a row. On a geological map they are multiple broken slivers in harlequin colours.

Dufton Pike and its neighbours, Knock Pike and Murton Pike, are distinctive pointed hills which contrast with the horizontal layers of the Pennine plateau behind. The hills are part of the Cross Fell 'inlier', a 25 by 2 kilometre area of older rocks peeping through younger Carboniferous and Triassic strata. The inlier is made up of slices of the three major Lake District rocks: older Ordovician sedimentary rocks, younger Ordovician volcanic rocks and Silurian sedimentary rocks, all broken up by geological faults. On a map they appear to be a detached part of the Lake District but below the surface they are connected to it. The volcanic rocks are lavas and tuffs (ejected ash and rock fragments). While their origins are volcanic, disappointingly these conical Pikes are not volcanoes. They stand out as steep-sided peaks simply because of movements and dislocations in the Earth's crust and because all the Pikes are harder than the surrounding rocks.

Older Cumbrian rocks like these have experienced three major episodes of mountain building and each of these events re-activated old fault lines. Mountain building (called an orogeny) happens at continental scales over millions of years. The first episode called the Caledonian Orogeny was between 490 and 390 million years ago. The second, the Variscan Orogeny, affected Britain around 290 million years ago. The third, the Alpine Orogeny, began 65 million years ago and is still happening now. The Alps and Himalayas are still growing despite being eroded by frost and ice. The notorious Helm Wind (the only named British wind and a strong northeasterly) blows down the escarpment here. Wheatear and buzzards are a common sight and near the summit are areas of bilberry (help yourself).

There is parking in Dufton village. The Pike is a 6 kilometre return hike [NY700266]

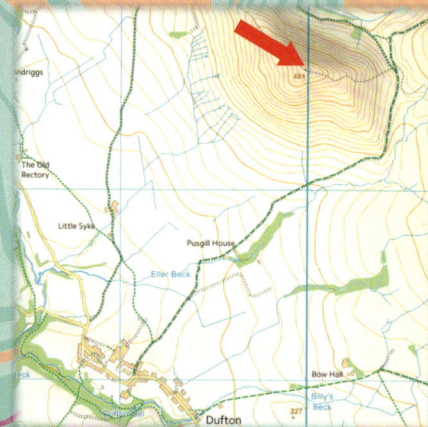

Knock

Dufton

Brampton

Murton

Dufton Pike and Murton Pike viewed from Knock

28

Hartside

Hartside is a cyclist's nemesis. After meandering gently across the Vale of Eden, the A686 road out of Melmerby climbs abruptly to over 750 metres above sea level.

Why the dramatic change in slope? When you get to the top look north and south, you can see clearly for miles just how sharply the Pennine escarpment rises from the softly undulating plain. The journey from Melmerby took you across one of the most visible landscape-scale rock dislocations in England: the Pennine Fault System. Although the Fault and its history are complex, put simply it is a break in the Earth's crust which separates younger rocks in the Vale to the west from the older rocks of the Pennines to the east. Most of the Vale's rocks are 250-million-year-old red desert sandstones of the Permian and Triassic periods. The rocks of the Pennines are largely 330-million-year-old Carboniferous, (but east of Appleby there are even older Silurian and Ordovician strata).

This dramatic displacement (around 600 metres vertically) continues deep into the Earth and probably has its origins in the building of an ancient mountain chain more than 400 million years ago. But it experienced a major re-activation around 250 million years ago when the Fault was a strong influence on the deposition of the red sandstones of the Eden valley. It was active again around 65 million years ago, when the Alpine mountains began to form and there is evidence that it continues to be an area of weakness in the crust. On 9 August 1970 an earthquake with an unusually high magnitude for the UK (4.9 on the Richter Scale) had its epicentre near the southern end of the Pennine escarpment. The numerous twists on the ascent from Melmerby navigate through a series of glacial meltwater channels. As the last ice sheet was melting its surface slowly got lower and the meltwater coursing along the successive ice sheet margins, carved what are now mainly dry channels.

Bike (or drive) from Penrith or Alston. It's uphill both ways [NY646418]

View of the Pennine escarpment from south of Melmerby

29

Mosedale

There are a few Mosedales in Cumbria, perhaps not surprising as it means 'bog in a valley'. This one in the northern fells is the source of the River Caldew.

Mosedale is a beautiful, glaciated valley with a typical 'U' shape. For the most part it cuts through 470-million-year-old crushed, bent and broken Ordovician mudstones and siltstones. But glaciers and then the river have cut down deep enough to reveal something else beneath: a younger granite plus evidence of how this granite 'baked' the rocks around it. The Skiddaw Granite doesn't show a lot of itself at surface, but the intense heat from when it was molten, around 405 million years ago, completely transformed the older mudstones and siltstones around it. The alteration caused by the baking has a concentric pattern. The zone closest to the granite is a hard slate filled with garnets, while further out there are crystals and spots of other minerals (chiastolite) on the slate surface. This process of change through heat (and/or pressure) is called metamorphism and the concentric zones are called a metamorphic aureole.

The Skiddaw Granite that brought the heat is quite different to the slatey rocks around it. It is made of interlocking crystals of black mica, white feldspar and opaque quartz that solidified in a chamber of molten rock around eight kilometres below the surface. The granite changes the further north you go; as the magma cooled, hot fluids altered its minerals so that it became a rock called greisen. These hot fluids were mineral-rich; one reason why the rare and valuable metal tungsten was once mined in the valley.

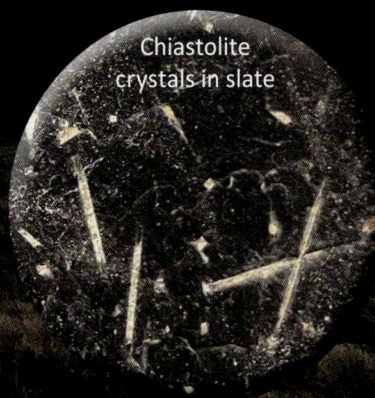

Chiastolite crystals in slate

Drive up the valley from Mosedale hamlet. There is parking beside the road [NY327327]

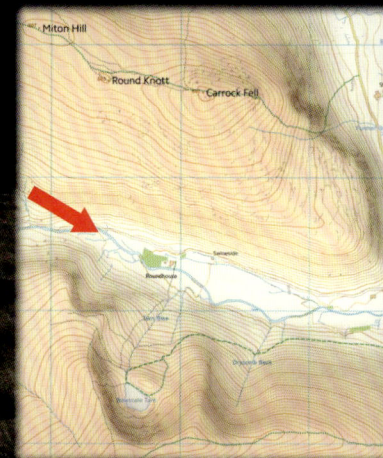

Skiddaw Granite altered to greisen in the bed of the River Caldew

30

Penton Linns

Liddel Water forms a part of our border with Scotland. Downstream of Penton Bridge the river cuts through a series of dramatically bent and upended rocks.

The prominent rocks are limestones but they are within a sequence of sandstones, mudstones and coals; the pattern is repeated many times. This variation in rocks is a result of the different environments that existed one after the other 330 million years ago. Fossil shells and coral tell geologists the limestones were deposited in shallow, warm seas. The shales were mud in lakes and coastal lagoons; the sandstones, sand in rivers and deltas; the coals were once vegetation that grew in swamps on the river plains and rotted and compressed to become peat and then coal. All the rocks are from the Carboniferous Period. Just downstream from the bridge the rocks have been bent into an arch (called an anticline). Below this the angle of the rocks steepens. They are almost vertical and look like tombstones; they divert the course of the Liddel. All these rocks were once horizontal, but pressures within the Earth during successive periods of mountain building have caused the bizarre geometry they have now. Even rocks can be bent if you do it slowly enough and bury them deeply.

Like many Northern oak woodlands, Penton Woods bursts into life in spring with carpets of bluebells and white-flowered wild garlic – a heady combination of colours and scent. Later, as the early flowers fade and wither, the wood is dominated by the dense bracken undergrowth.

Just beside the bridge on the Scottish side there is a small parking place. Take the path downstream [NY433775]

Almost vertical beds of Carboniferous limestone in the Liddel Water

31

Shap Summit

The rocks near Shap Summit on the A6 may look just as peaceful as the road is since the M6 was built. But, like vehicles attempting the infamous climb 60 plus years ago, these rocks have known tough times too.

You might want to drive as slowly as they did back then to see how buckled and broken these roadside rocks are. They look every one of their 425 million years. They are slates and sandstones from the Silurian Period and were once mud and sand that settled at the bottom of an ocean. They were buried under perhaps 8,000 metres of sediment. The pressure of burial at that depth makes rock capable of being warped and shattered. It was a long period of instability and mountain building in the Earth's crust around 400 million years ago that did the damage. Take a closer look at the rocks and you will see that many are almost vertical and some have been folded and virtually corrugated. All of them have a layered texture (geologists would say 'foliation' or 'fabric'). These layers are nothing to do with their original horizontal deposition as sediment in the ocean; this is yet more evidence of the huge pressures their constituent minerals have endured.

How come something that was buried eight kilometres deep is at the Earth's surface today? The answer is millions and millions of years of erosion and a little more mountain building. That's hard to believe but then so is the fact that until the 1960's attempting to take a car up and over Shap Summit was regarded by drivers as a tough challenge, with the expectation that you'd have to pull in at the top to let the engine cool down. Ask your grandad!

THIS MEMORIAL PAYS TRIBUTE TO THE DRIVERS AND CREWS OF VEHICLES THAT MADE POSSIBLE THE SOCIAL AND COMMERCIAL LINKS BETWEEN NORTH AND SOUTH ON THIS OLD AND DIFFICULT ROUTE OVER SHAP FELL BEFORE THE OPENING OF THE M6 MOTORWAY.

REMEMBERED TOO ARE THOSE WHO BUILT AND MAINTAINED THE ROAD AND THE GENERATIONS OF LOCAL PEOPLE WHO GAVE FREELY OF FOOD AND SHELTER TO STRANDED TRAVELLERS IN BAD WEATHER.

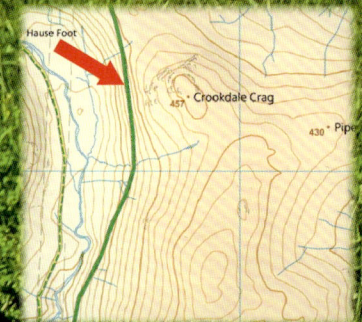

The A6 is not busy but traffic is fast. Park near the memorial [NY555053

Folded and faulted Silurian rocks beside the A6 just south of Shap Summit

32

Tarn Hows

Tarn Hows has been a well-known beauty spot since the Victorian era. Geologists like it here too, because it's somewhere they can study a very significant change in Lake District geology.

Tarn Hows is one of the places where there is a transition from what was once an ancient volcanic land above sea level, to a deep ocean called the Iapetus. Along the southeast shore of the Tarn the volcanic rocks that form the central fells, disappear beneath younger limestones, siltstones, sandstones and mudstones. These oceanic sedimentary rocks are mostly Silurian, that is between 420 and 440 million years old. Initially the water was shallow and the limestones were deposited. Over millions of years the ancient ocean deepened and sand, silt and mud more than six kilometres thick covered its floor. The two continents which bordered the Iapetus Ocean collided and closed it. The shape of the Lakeland landscape reflects these events; the sedimentary rocks around Coniston and Windermere are far less rugged and not as high as the more resistant volcanic rocks of the fells to the north. Just 200 metres or so northeast of the tarn the layers of limestone rocks you see are no longer horizontal, they are inclined at around 60 degrees. They also show regular cleavage at an angle completely different to that. The cleavage and slope in the rocks are because the original sediments have been compacted and then tilted by pressures deep underground.

Beatrix Potter lived close to Tarn Hows and she used to own part of it before she presented it to the National Trust. It is a justifiably popular place, with superb views of the fells and excellent access to the tarn. But you only have to walk a little way off the beaten track to seek out these ancient rocks and you will have the place to yourself. Curiosity about geology often reveals secret places.

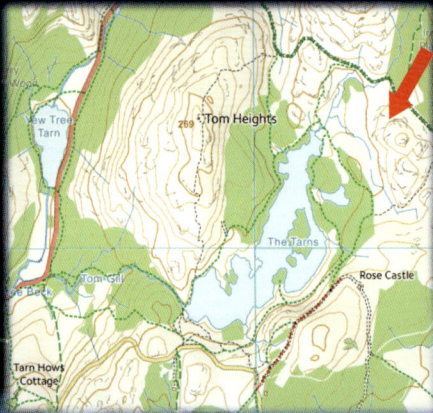

There is parking and a good footpath around the Tarn [SD331999]

Tilted and deformed Silurian limestones northeast of Tarn Hows

33

Trusmadoor

Tucked away in the northern fells is a beautiful little valley with a unique name and two rock stories to tell......millions of years apart.

Trusmadoor - the first two syllables of its name are rare – *Trus* and *ma* are from an ancient Celtic language called Cumbric, used in the kingdoms of Cumbria after the Romans. The whole word is literal, it means place of the mountain gap to which the English word 'door' has been added.

The rocks in the crags on the valley side are slates. They are from the Skiddaw Group which makes up most of the northern Lake District and of course the mountain they are named after, Skiddaw itself. Their formal geological name used to be Skiddaw Slates but not all the rocks are slate so geologists changed the name. During the Ordovician these rocks were mud, silt and sand in a deep ocean. Millions of years of being buried and heated deep in the Earth, and then uplifted, broken up and eroded, has produced the rocks and the rounded hills we see today. Life on Earth was primitive 480 million years ago and existed mainly in the sea. Fossils from some of those primeval animals, called graptolites, have been found in the slates at Trusmadoor. The way graptolites changed their shape as they evolved has helped palaeontologists to unravel the sequence and development of life on our planet back then.

Move forward to just 15,000 years ago. Glaciers which had occupied Lake District valleys for many thousands of years were melting; torrents of water and debris draining under and out of the ice cut channels in the bedrock. Trusmadoor is one of those Ice Age channels, left high, dry and hanging now. Alfred Wainwright described Trusmadoor more dramatically 'a lonely passage between the hills....a natural railway cutting. What a place for an ambush and a massacre'.

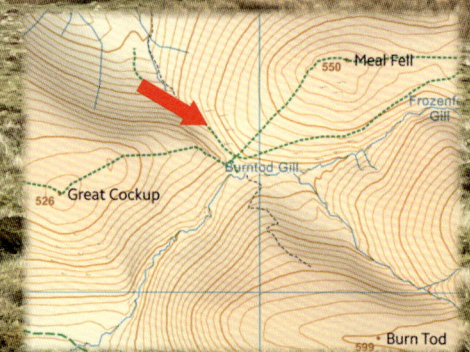

A 7 kilometre return hike up Longlands Beck or 12 kilometres across Great Sca Fell from Fellside [NY279335]

Slates from the Skiddaw Group of rocks on the side of Trusmadoor

34

Brampton

Anyone who has driven the A69 past Brampton will recognise how hummocky, sandy and red the local landscape is.

Stretching southwards from Brampton to Cumrew is an undulating expanse of sand, gravel, silt and clay around 44 square kilometres in area. Geologists named it the 'Brampton Kame Belt'. 15,000 years ago rivers and lakes draining England's last stagnating ice sheet deposited one of the largest areas of glacial meltwater sediments anywhere in the UK. Only a few thousand years earlier an ice sheet over 1000 metres thick, had advanced down the Irish Sea, across the Solway Plain, joining with ice from the Lake District before flowing south and eastwards through the Eden and Tyne valleys. As the climate warmed, the ice progressively wasted and receded southwards and westwards and the meltwater transported and sorted the debris the ice had carried. The debris (and thus the soil) is red because the ice sheets had eroded the local Triassic sandstone bedrock.

By looking at the shape and composition of this undulating landscape, and comparing it with places in the world where ice sheets are melting today, we can understand how the different ridges, humps and hollows formed. What were once channels for torrents of meltwater are now quiet, dry valleys. Ridges, like the one in Brampton itself, were rivers carrying debris under, on top of, and within the glacier. Flat-topped hills were ice-walled lakes into which deltas flowed, depositing sand, and often delicately laminated silt and clay. Enclosed hollows are 'kettle holes', places where large blocks of dead ice were buried by sediment and then melted, leaving a depression. Talkin Tarn, a Country Park, is in one of the largest depressions. Bog bilberry, very uncommon south of the Scottish Border, grows in woodland here. The glacial deposits have left a fertile landscape, but its ups and downs mean it is often left to pasture. Cattle and sheep farming dominates the landscape but the River Gelt and smaller streams like the Cairn Beck have cut valleys through these soft sediments and are home to ancient woodlands and diverse wildlife.

Brampton has parking and is served by bus; try the footpath along the Brampton Ridge [NY536615]

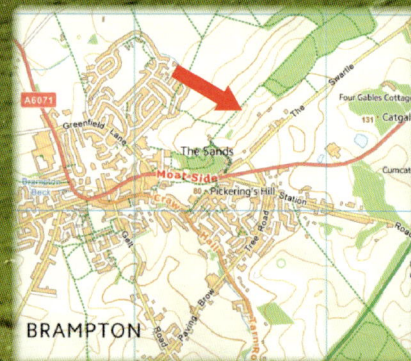

BRAMPTON

Typically undulating landscape of the 'Brampton Kame Belt' east of Faugh

35

Glasson Moss

Surrounding the Solway coast are some of the largest lowland raised bogs in England. Glasson Moss is one of these bogs.

Raised bogs, mosses or mires are areas of peat that started to develop after the last ice sheet melted away around 11,000 years ago. The bogs occupy poorly drained depressions and flat surfaces and are sustained by rainwater. Plants like sphagnum grow in these acid conditions. The vegetation progressively builds up to form raised mounds of peat which can reach several metres high. Bogs require permanently waterlogged conditions for their plants to thrive and accumulate and it is that lack of oxygen that inhibits the work of microscopic organisms, so the vegetation only partially decomposes.

At Glasson Moss the peat reaches thicknesses of over ten metres. It was once the site of extensive peat extraction, most of which went for horticulture. In recent years restoration projects are encouraging its recovery. It is now a nature reserve and boardwalks and a viewing tower mean you can appreciate the extent of the Moss and see the rich wildlife too.

Nearby are other mires, Wedholme Flow, Bowness, Drumburgh and Finglandrigg, and towards the border with Scotland, Solway Moss. In 1771, after a night of torrential rain, Solway Moss burst and covered 400 acres with 30 feet of peat and mud. It destroyed homes and left 35 families destitute. Bogs are fragile places and face many threats. Changes to water levels through agriculture, tree planting and worst, peat extraction, have had disastrous effects. But even activities such as grazing, burning, or our leisure pursuits, can damage them. Peat bogs are places where huge volumes of carbon are stored naturally; maintaining and restoring them helps in the fight against our overheating planet.

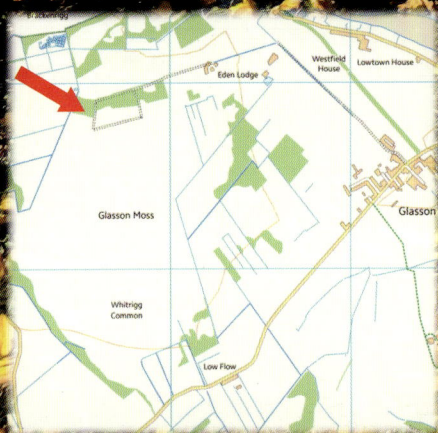

Take the road into the caravan park and then the track to the viewing tower [NY236609]

The fault is great in man or woman who steals a goose from off the Common.

But what can plead that man's excuse who steals the Common from the goose?

Looking south over Glasson Moss from the viewing tower

36

Glenridding

In December 2015 a reporter from a national TV station described the scene as floodwater, carrying rocks, boulders and silt, overwhelmed Glenridding. 'Unprecedented!' he declared. It was devastating, yes, but not unprecedented.

Like all mountainous areas the Lake District is an active landscape. It may look stable and peaceful but its steep slopes and narrow valleys means it is susceptible to extreme weather. Increases in heavy rainfall as our climate changes will inevitably have more impact. Geological maps can be used to identify where these events have taken place in the past. Look at the map of Ullswater: areas of sand and gravel ('alluvial fans') spread into the lake at the mouths of mountain streams. The village of Glenridding sits on one of these; the fan, now a grass-covered expanse of flood debris can be plainly seen from Place Fell opposite. These fans started to build as the glaciers began to recede 15,000 years ago and every large storm since has added to them. Such events are not unprecedented, merely beyond living memory and they will happen again.

Spreads of alluvium provide rare areas of flat, fertile ground and it is natural that humans have occupied them wherever they occur. The village of Buttermere sits on the alluvial fan produced by Mill Beck flowing down from Newlands Hause. The fan has reached to the other side of the valley and in doing so created two lakes, Buttermere and Crummock Water.

Living in a landscape that will become more active in a changing climate presents challenges. Investing in flood defence walls and dredging stream beds may be one answer. In other active landscapes, like the eroding coast of Norfolk, the scale of the challenge is greater and the inevitably unpalatable solution for some communities is not to defend but to adapt, and sometimes to recognise the unassailable force of nature, and retreat.

There is parking in the village. Walking to the top of Place Fell is a strenuous 7 kilometre return hike from Patterdale [NY387169]

View of Glenridding and the alluvial fan from Place Fell

37

Great Asby Scar

Limestone pavements are rare in Britain and cover less than 2000 hectares in total. The best examples are in Cumbria and North Yorkshire. Those at Great Asby Scar are spectacular.

Limestone pavements are areas of bedrock which have been scraped bare by ice sheets. Limestone is a soluble rock and it has a natural system of grid-like joints and cracks. Water follows these and then dissolves the surface into blocks (called clints), separated by fissures (called grikes). While this natural 'pavement' is probably less than 15,000 years old, the limestone rock is much older. It was formed in a Carboniferous sea 335 million years ago. In some places you can see the fossils of corals in the rock, evidence that this sea was warm, shallow and clear and that at the time, what is now England, was very near the Equator. The rocks at Great Asby Scar are special and have given their name to rocks of the same geological age all over the world; they are called Asbian.

The plants which grow on limestone pavements take advantage of the different habitats the clints and grikes provide. Hart's-tongue and rigid buckler ferns and dog's mercury grow in the shadier grikes. These pavements are a very sensitive environment and pressure from grazing can seriously affect them; fortunately, a large part of the area is now a National Nature Reserve.

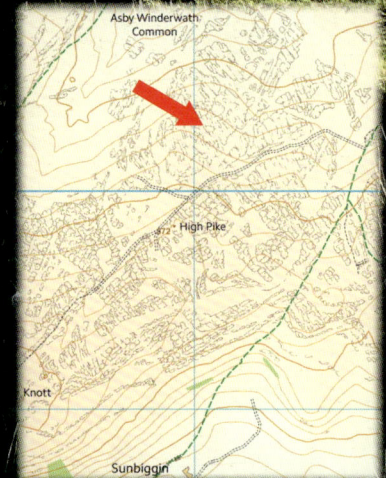

A 4 kilometre walk eastwards from Orton village [NY654096]

Limestone pavement at Great Asby Scar

38

Hale Moss

Going deep into caves is best left to cave explorers, but at Hale Moss you can peep into the entrance of one from the roadside.

Natural caves can occur in lots of different rocks but most caves form when the rock can be dissolved by water. There are caves in Cumbria where there is gypsum, as it dissolves very easily and quickly, but the most common rock that has caves is limestone. Limestone was once limey mud and the shell debris of billions of fossils and is, essentially, calcium carbonate. Limestone rocks also have lots of cracks and joints in them and that provides a pathway for underground streams which progressively dissolve the rock away. South Cumbria has a lot of thick limestones that slope downwards and as a result a lot of caves. Precisely how did they form? Rain absorbs carbon dioxide from the air, and also from organic matter in the soil and as the water percolates through it becomes a weak acid dissolving the rock.

Hale Moss caves, like the others in this part of Cumbria, are in one of the thick Carboniferous Limestones (here the Urswick Limestone). This limestone is around 335 million years old and 120 metres thick. The cave is over 200 metres long and at the entrance you can see it has a 'slot' like shape, evidence that this part of the cave was dissolved by water flowing above the water table. Caves eroded beneath the water table are usually more rounded in shape. When was the cave formed? Probably in the last 500,000 years at times when groundwater levels were much higher than they are now. This complex of limestone caves is within diverse woodland. Patches of species-rich limestone grassland also occur within the site and add greatly to the wildlife interest.

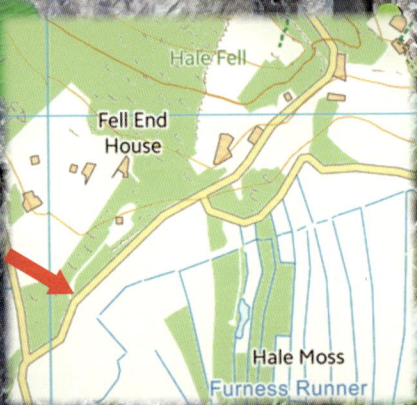

The caves are on the northern verge of a minor road southwest of Hale [SD501777]

Cave entrance at Hale Moss

39

Helvellyn

This is not so much about the top of England's third highest mountain, more about the ridges that lead to its summit and the tarn that separates them.

Climb all 950 metres of Helvellyn from the east and you will need to scramble along one of two sharp ridges: Striding Edge or Swirral Edge. Both are arêtes, the narrow razor of rock left when glaciers carve out the valleys on either side. The freezing and thawing of snow and ice, plus gravity, then sharpen the ridges further. In between these two Edges, lying in a deep bowl, is Red Tarn. In Scotland the bowl would be a corrie, in Wales a cwm, in France a cirque. Red Tarn is typical, it has three steep sides and a low ridge across its fourth side. The bowl it sits in was carved out by snow that turned to ice, which then progressively eroded and deepened the hollow. Arêtes and cirques are characteristic erosional features of high mountain landscapes everywhere. The rocks that form the bedrock of Helvellyn are volcanic ash and rock fragments which were erupted 450 million years ago. The glaciers which eroded the arêtes and cirques are much younger, less than 2.6 million years old. There have been several Ice Ages during that time; the last major one melted around 15,000 years ago. Then around 12,000 years ago there was a short cold period and small ice caps and glaciers formed again in the Lake District's high mountains. All these events have cumulatively helped to sculpt the dramatic landscape that we enjoy today.

The crags at the back of the Red Tarn cirque are one of Cumbria's botanical hotspots, with many montane plants. They include rose-root, mountain sorrel, moss campion and purple saxifrage – particularly where cracks in the rocks have become filled with calcite (calcium carbonate). Helvellyn is probably a Celtic (Cumbric) name meaning pale yellow upland. That sounds right as the top is a flat grassy plateau and not the sharpest summit in the Lake District. But its razor ridges and the panoramic views from its summit make for a great hike.

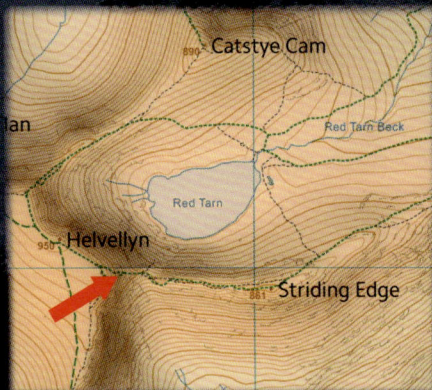

The summit is a strenuous 7 kilometre return hike from the car parks south of Thirlspot [NY343151]

Looking down on Striding Edge and Red Tarn

Inner Solway

40

Seen from space there are two features that dominate Cumbria's geography. One is the Lake District and the other is the Solway Firth.

Unlike its prominent neighbour to the south, the Solway may appear passive; a flat land devoid of much appeal, and certainly any geological interest. But to adapt a quote about the land around The Wash 'Any fool can appreciate mountains, it takes someone of true discernment to appreciate the Solway'.

The inner Solway is perhaps one of Cumbria's most active environments. The estuarine system of mudbanks, sandflats and saltmarshes is dynamic, with shifting channels and multiple episodes of erosion and deposition. This place is one of the youngest 'rocks' in our book and it continues to change dramatically. In the late 19th century tidal flows and then thick ice moved at such speed that they damaged the iron piers of the Solway rail viaduct near Bowness-on-Solway, a precursor to its ultimate demise. Within a decade of 1880 the channel to the harbour at Port Carlisle shifted, leaving the harbour and canal to the City of Carlisle, silted up and marooned.

The foreshore muds and sands change their distribution daily and tides can strip them bare to reveal scaurs (or scars); banks of stones and clay. These were deposited by ice sheets around 20,000 years ago, when the Solway was a junction for glaciers flowing across northern Britain. On occasions the stumps of a submerged forest, around 7000 years old, are revealed. Today the Solway is of international importance, to wildlife and to naturalists. It is home for migrating, wading and wintering birds and is the third largest intertidal habitat in Britain. Tranquil it may be, passive it is not.

Park at the west end of the village and walk the shoreline [NY220627]

Remains of the rail bridge across the estuary west of Bowness-on-Solway

41

Mickleden

Most Lake District valleys have little hillocks on their floors and sides. Mickleden in upper Langdale is typical. These humps and bumps are moraines.

Glaciers scrape off and collect a lot of stuff. When they stagnate, retreat and melt they leave that debris behind. It's called moraine. The debris can be everything from boulders, to sand and silt and clay. There are lots of types of moraines, usually classified by their shape and position. When these are compared to active glaciers today, geologists can work out how and when they formed. In the valley, near where Stake Gill and Rossett Gill meet, there are many ridge-like mounds of sand, gravel and boulders, some of which have been cut through by the streams. While these can look disorganised, detailed air photography plus pollen and cosmic ray dating techniques can make sense of them. We believe these are lateral and terminal moraines built up by meltwater at the front and sides of a retreating glacier. There is more hummocky ground in Langdale Combe higher up, where the Cumbria Way follows Stake Pass. These are moraines too, but altered by erosion and slumping down the steep slopes since the ice left.

There have been recent studies on both sets of moraines; all part of wider research to understand exactly how the last Ice Age that covered Britain behaved. There is much debate about these features but geologists agree that ice was covering the high plateaus and valleys of Cumbria during a brief very cold episode around 12,000 years ago. If we want to better understand how climate change will affect our planet in the future, this research is more than academic. A famous geologist, James Hutton, once said 'the present is the key to the past' but the past can also be the key to the future.

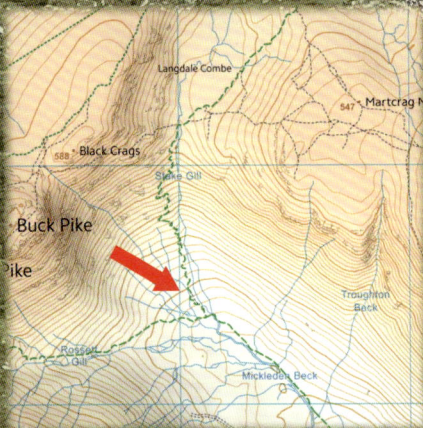

A 3 kilometre level walk from Old Dungeon Ghyll to the confluence of Stake Gill and Mickleden Beck [NY261073]

Glacial moraines at Mickleden in the Langdale valley

42

Ravenglass

The intricate spits and dunes of Drigg and Eskmeals guard Ravenglass and its estuary which is fed by three rivers.

The rivers Irt, Mite and Esk drain a huge area of southwest Lakeland, including the Sca Fell range. Their confluence is in the estuary west of Ravenglass village. The valleys of these three rivers, over-deepened by glaciers, once used to run separately to a coastline more than a kilometre to the west but were submerged as sea levels re-bounded when ice sheets melted around 15,000 years ago. The sea then re-worked the glacial debris and built up the shingle spits of Drigg and Eskmeals. The longshore drift of sediment is to the south at Drigg and to the north at Eskmeals. In the last few thousand years onshore winds have created extensive sand dunes here. Dunes are also found to the south at Sandscale, and to the north at Mawbray and Silloth. Recent research shows that today Ravenglass estuary, while dynamic, is in equilibrium; flood and ebb tides import and export similar amounts of sediment. How the estuary will respond to predicted global rises in sea level remains unknown, but current planning policy assesses the risk to human infrastructure as being of 'low impact'.

Ravenglass has a rich human history too. In the dunes Neolithic and Bronze Age flints and artefacts have been found. In Roman times the estuary mouth was wider and they maintained a port here, and also a fort called Itunocelum, plus a large bath house complex. Up to the Industrial Revolution Ravenglass continued to be a significant port, shipping iron, copper, slate and granite; all mined and quarried in the surrounding fells. Today the old mineral railway line, La'al Ratty, is a popular destination for tourists taking the ride into Eskdale. The dunes have abundant wildlife, including many hundreds of plant species, wading birds, wildfowl, insects and rare natterjack toads; this is an internationally important habitat.

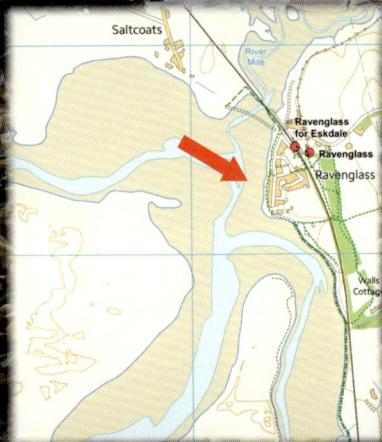

Ravenglass is a stop on the Carlisle-Barrow train line and on the Ratty railway [SD084964]

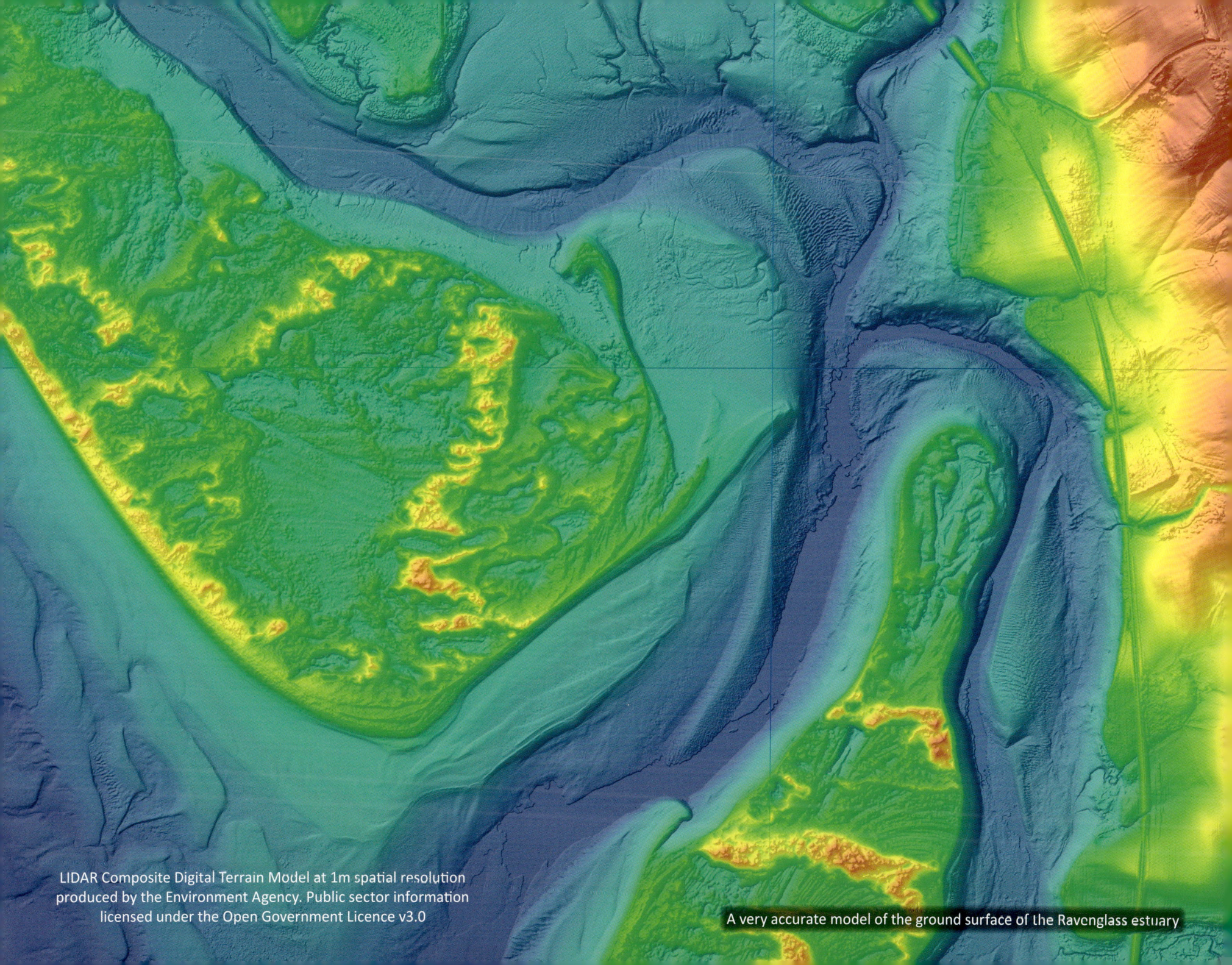

A very accurate model of the ground surface of the Ravenglass estuary

43

Skelsmergh

Just north of Kendal the A6 weaves through a collection of whaleback mounds. They are called drumlins.

On the lower ground of Cumbria overlooked by the mountains of the Lake District and the Pennines, drumlins are everywhere. Images from space show the vales and plains filled with what look like thousands of giant half-buried eggs. They are all aligned, and they appear to 'flow'. Drumlins are a rich area of study for geologists but exactly how and when they formed remains a mystery. All agree they formed under an ice sheet and that they show the direction of ice flow, but beyond that their origin is a debate that has rolled on for decades. The widely accepted theory is that ice sheets ride over and deform a whole variety of glacial debris beneath them and streamline it. But until someone is able to watch a drumlin form underneath an ice sheet, exactly how they are constructed remains theoretical! Drumlins and other glacial features have been used to try to work out how and when ice sheets moved across Cumbria. Satellite images, aerial photography and recently radar and laser technology (bouncing beams from a plane to the ground to precisely measure the surface height) have helped scientists enormously. That research confirms that the Lake District had its own ice cap and also suggests that ice sheets flowing from there, the Irish Sea, and the Vale of Eden have come together and, at times, completely switched direction. The result is drumlins overprinted on drumlins; 20,000 years ago the Cumbrian lowlands were a real mixing bowl.

Our undulating landscape is one piece of evidence of the Ice Age, another is well known to many gardeners across Cumbria: boulder clay (geologists now call it till or diamict). The mud, sand and stones that were once on, in, beneath and churned up by the ice are now the sticky, stony clay that makes digging such hard work.

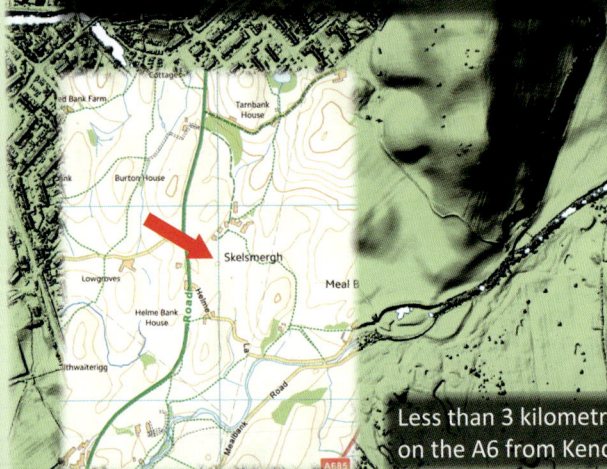

Less than 3 kilometres drive north on the A6 from Kendal [SD529952]

LIDAR Composite Digital Terrain Model at 1m spatial resolution produced by the Environment Agency. Public sector information licensed under the Open Government Licence v3.0

View southwest over drumlins from the A6 near Skelsmergh

44

Slater Bridge

A few metres north of Slater Bridge in Little Langdale is a rocky knoll. Its western end is a gentle slope of bare, smoothed rock. Over the wall at the eastern end is a little crag.

Look closer and you see the smooth surface is scratched, a fact much appreciated by geologists, as unlike Antiques Roadshow experts, rock men like a little wear and tear. Rocks like this help us to understand what our landscape has experienced in the past. This rock is composed of 450-million-year-old volcanic material, but its shape and the smooth and scratched surface are much more recent. These features are evidence that this place was once under a glacier several hundred metres thick which moved down the valley scraping and grinding everything beneath it. The scratches on it are called striations, they are caused by grit and stones embedded in the base of the ice steadily sanding down the bedrock. The gentle western slope and the crag at the eastern side show how the glacier rode over the hill and then "plucked" the rock at its lee end. Rocks with this shape are called roche moutonnée. Not because they resemble sheep but because they look like an 18th century Frenchman's wig, which was apparently smeared with mutton fat to make it sleek. Rather them than me.

You can find rocks like this all over the Lakes, they and features like 'U' shaped valleys and moraines are clues to our frozen past 20,000 years ago. Slater Bridge has been described as the most beautiful bridge in the Lake District. It is a delicate stone arch and a flat span made of a single thin slab of slate, both connected by another glacially smoothed rock. The bridge was constructed by quarrymen 300 years ago to shorten their journey to work.

600 metres walk south from Little Langdale [NY312030]

Direction of ice flow

Glacially smoothed bedrock north of Slater Bridge

45

Wastwater

It is England's deepest lake and in 2021 the Royal Geographical Society named it as one of the seven natural wonders of Britain.

When you enter Wasdale and get your first glimpse of the lake it is hard not to be over-awed. Wastwater's long straight valley draws in your eye. In the distance are the mountains of Yewbarrow and Kirk Fell and bordering the southeast side of the lake are the most dramatic screes in Cumbria. The valley is almost five kilometres long and the bottom of the lake is 79 metres below its surface. The valley was scoured and over-deepened by glaciers. The last one retreated only about 15,000 years ago. But it is the screes that dominate the near view. Vast cones of angular rocks and boulders, at least 200 metres high, cascade down from rocky gullies and crags into the lake. The crags are 450-million-year-old Ordovician lavas. The debris (talus) of the screes is the same rock. The screes began to form fully as the glacier, which filled the valley, melted away and freezing and thawing shattered the bedrock. It is probable the valley sides also 'de-stressed' and slid away after billions of tons of glacial ice disappeared. The scree slopes are steep and you can pick out by the lack of vegetation where they continue to move through gravity and water flow, especially after periods of heavy rainfall.

You get a great view of the screes from the north-western shore of Wastwater, but if you are tempted to see them close up, please take care. In places the lakeside path is very difficult to follow and the huge boulders and gaps between them make the screes a dangerous place for the ill-prepared and a regular venue for the mountain rescue team.

Easiest viewed from the road on the northwest side of Wastwater [NY151044]

Wastwater and its screes from the northwest

46

Florence Mine

Two rocks, coal and iron ore, were the backbone of the economy of west and south Cumbria for more than 100 years. Florence Mine was a major player in that economy.

Huge iron ore (hematite) deposits were known to exist near Egremont. Florence Mine was opened in 1914 to extract them. When large scale production ended in the 1970's over 100 million tons of ore had been mined. There were several other large hematite deposits in the area and these were mined at Beckermet and Ullcoats, as well as Frizington, Cleator Moor, Millom and Barrow. Iron ore mining in Cumbria, which began in Roman times, stopped completely in 2007 when British Nuclear Fuels at Sellafield ceased to need cooling water from the mines and stopped paying for the pumping which kept the mines dry and workable. Hematite (or haematite) formed when warm iron-rich fluids percolated through rocks and replaced areas of Carboniferous limestone. The ore tends to occur near fractures (faults) in the rocks. Hematite in west Cumbria commonly forms bizarre organic shapes and is sometimes called kidney ore because of its 'botryoidal' crystal structure. The mine is famous for the spectacular minerals it has provided to museums around the world and it is now a Site of Special Scientific Interest.

Anyone who has visited a hematite mine will know just how red and mucky they are so working in these mines was not a subject for the Romantic Cumbrian poets. But the poet Norman Nicholson, thought differently. Nicholson lived all his life in south Cumbria amongst the iron mines and fells. He had a deep appreciation of industry, geology and the environment. His poems about them are as apposite as they are eloquent and down to Earth.

The mine is now an arts centre southeast of Egremont [NY170103]

Hematite from Florence Mine, image by kind permission © The Trustees of the Natural History Museum, London

Old buildings and shaft at Florence Mine

47

Furness Abbey

On the outskirts of Barrow is Furness Abbey, a 12th century monastic settlement and once the richest in northwest England. It may now be a ruin, but the quality and sophistication of the architecture are breath-taking.

The Abbey is still the largest structure in Cumbria that is built of red Triassic sandstones. Not to be outdone, the Victorians constructed the monumental 19th century Barrow Town Hall of the same red stone. The quarries for the Abbey were said to be in the adjacent Vale of Nightshade; the stone for the Town Hall was local too, it came from Hawcoat. The building stones are from a set of rocks known as the St Bees Sandstone and their composition and colour is a product of the way they were formed – as sand in rivers which flowed across a hot desert plain 250 million years ago.

People have always used local resources for building and the incredible diversity of Cumbria's rocks is reflected in the variety of stone used in its buildings, from prehistoric and Roman times to today. The red Triassic sandstones used here are common elsewhere in lowland Cumbria. In the Lake District the houses are of Ordovician and Silurian slate and volcanic rocks. In the Pennines the farms and walls are of brown and grey Carboniferous sandstone and limestone. Even where there is no solid rock, people have used stones, rounded and carried from afar by ice and water, to build the boulder and cobble houses of the Solway Plain. The range of stone buildings in the county is as wonderful as it is unique.

2 kilometres northeast of Barrow town centre [SD217718]

St Bees Sandstone of Furness Abbey

48

Gilsland Spa

In the late 18th and early 19th centuries, before going to the seaside became popular, people went to a spa for their health. Two springs, one sulphurous, the other iron, in the valley of the River Irthing near Gilsland, attracted locals and visitors alike.

Bath houses once stood beside the river but all that remains of that early tourist trade today is a re-built stone well edifice. It has a modern basin and pipe that dribbles clear water which smells strongly of rotten eggs and turns white when exposed to the air. 200 years ago people came to 'take these waters', drinking a glass or two a day and immersing themselves in it. According to Augustus Bozzi Granville, in his 1840 guide to the Spas of England, Gilsland's spring water was every bit as good as Harrogate's, (but he thought the facilities and qualities of the local population were very 'so-so'). He did, however, appreciate the natural beauty of the Irthing gorge, as did Sir Walter Scott. It was here, in 1797, at the Popping Stone, that Scott is reputed to have successfully proposed to Miss Charlotte Carpenter. Sadly, not everyone respects the natural and literary history of this place; in 2021 the Popping Stone(s) were bulldozed aside so trees could be felled. They have been returned, but the site is not the same.

The relevance of geology to these places of pilgrimage? The groundwater that feeds the two springs percolates through Carboniferous shales in the cliffs above. The shales contain the minerals iron sulphide and iron sulphate, the water dissolves these and brings them to the surface via a geological fault. The Popping Stones have an equally prosaic explanation; they are Carboniferous sandstones, rocks that are common in the cliffs nearby, but their rounded shapes may owe something to the hand of man, as well as river erosion.

Park in the car park next to the hotel and walk down to the river [NY636678]. The Popping Stone is 600m upstream.

The Popping Stone beside the River Irthing at Gilsland

49

Hallbankgate

The countryside between Talkin and Tindale is peaceful today but the scars of a noisier, industrious, past are everywhere: old quarries, overgrown mine dumps and derelict limekilns.

Mines and quarries occur all across Cumbria, but here, on the very northern edge of the Pennines, the exploitation of the county's natural resources was comprehensive. Every rock type in the area was put to use; coal, limestone, brick shale and sandstone were extracted; even zinc ore was smelted. Connecting all these endeavours was a pioneering railway system, the first to use wrought iron rails and the last to see George Stephenson's Rocket haul a waggon in earnest.

One hundred years ago and more the basis of the bustling local economy was its Carboniferous rocks. What were once, 330 million years ago, peat swamps, river deltas, muddy lagoons and coral seas had become coal, sandstone, shale and limestone. But it was coal that was the foundation for it all; a seam called the Little Limestone Coal. Not the thickest but so persistent it extends from Alnwick in Northumberland, to south of Alston. Villages and towns along its outcrop owe their existence to this understated 'little coal' and their communities their livelihoods. When this coal was a swampy forest, it must have been one of the most extensive in the north of England.

With a little help from us, nature is steadily recovering from all that industry. Trees are being replanted and wildlife is re-colonising the diggings, rail lines and moorland. Owls, barn, short-eared and long-eared, are commonly sighted and the higher ground is home to black grouse (blackcock), golden plover and merlin.

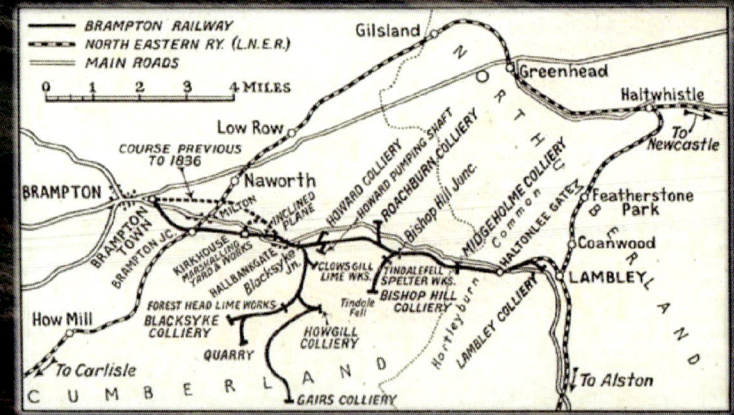

Rail map reproduced by kind permission © cumbria-railways.co.uk

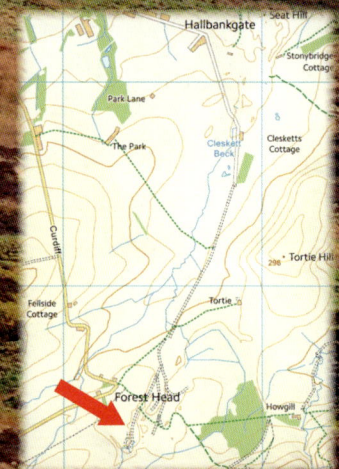

There is a small car park at Clesketts and you can walk across the old quarries [NY585574]. It's a 4 kilometre walk along the old rail line to Gairs mine.

View northeast over limestone and brickshale workings near Hallbankgate

50

Long Meg and her Daughters

'Next to Stonehenge it is beyond dispute the most notable relic that this or probably any other country contains.'

William Wordsworth, 1833

Quite the accolade for a stone circle that sits in a quiet corner of a distant northern county. But it is a very large and complex monument with an equally enigmatic narrative. 68 Daughters, plus Long Meg, stand around a perimeter almost 113 metres in diameter; and the stones are only part of the archaeology. The name? According to folklore, Long Meg was a witch, and her Daughters her coven; they danced on the Sabbath and were turned to stone. The stone circle is Neolithic and more than 5000 years old. Long Meg herself is a nine tonne block of 280-million-year-old Permian (Penrith) sandstone almost four metres high, which was probably quarried from cliffs beside the nearby River Eden. The majority of her Daughters are volcanic rocks from the Ordovician Borrowdale Volcanic Group of the Lake District, around 555 million years old. Most appear to be made up of angular fragments and blocks; breccias formed by the flow of volcanic lava and debris. In the last 2.6 million years ice sheets transported these very large stones (called erratics) from the Lake District to the Eden Valley. Given their large size it is quite possible the Neolithic community would have had to search for these stones from a wide area.

What the stone circle was for and why it is here are subjects of continuous archaeological debate. Such circles may have been important ritual sites and Long Meg is aligned to the winter solstice. It is probable that the people who erected it had a much greater affinity with the spiritual and material assets of the natural landscape than we do today. 5000 years ago such sites might have been places of celebration or gathering, monumental shrines to ancestors, or for religious ritual and it seems safe to assume that their purpose changed over time.

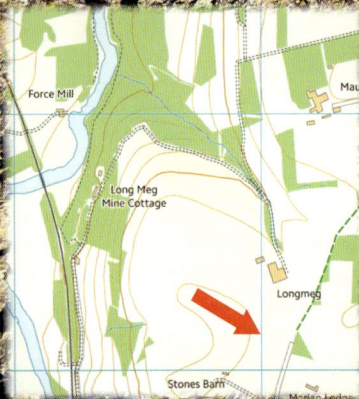

1 kilometre north of Little Salkeld. The road to the farm cuts through the stone circle [NY571372]

Long Meg and her Daughters

Long Meg Mine

51

270 million years ago Britain was only 15 degrees north of the Equator and the landscape in Cumbria resembled the arid coastal plains of the Persian Gulf.

Back then, water evaporated in shallow coastal lagoons and inland basins, leaving multiple layers of mud and salts. These sediments became the red shales and white gypsum and anhydrite beds that are found in the Eden Valley and along the west coast of Cumbria. Gypsum and anhydrite are both forms of calcium sulphate and they have been quarried and mined in these areas for years, producing the raw material for the plasterboard and chemical industries. Today only one active mine exists, at Kirkby Thore, and it is not easy to find a place where naturally occurring gypsum can be seen, mainly because it quickly dissolves away in water. But the dilapidated remains of one former mine, Long Meg, beside the River Eden, north of Little Salkeld, are still visible. The mine opened in 1880 and closed in 1976. Its underground workings are extensive and it produced more than five million tons of anhydrite. Beside the footpath which runs between the mine and the river to Lacy's Caves you can still find blocks of gypsum, although the minerals you see are secondary; they have been compressed, dissolved and recrystallised several times.

Native oak, birch and rowan woods fringe the river Eden, adding some stability to the steep banks and cliffs of red sandstone. In many places their exposed tangled-roots grow miraculously from thin soils and rock crevices with the help of their fungal root partners.

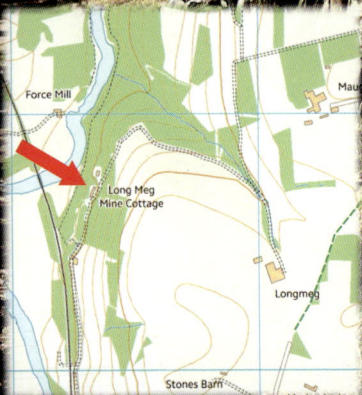

Take the footpath north from Little Salkeld for 2 kilometres [NY563378]

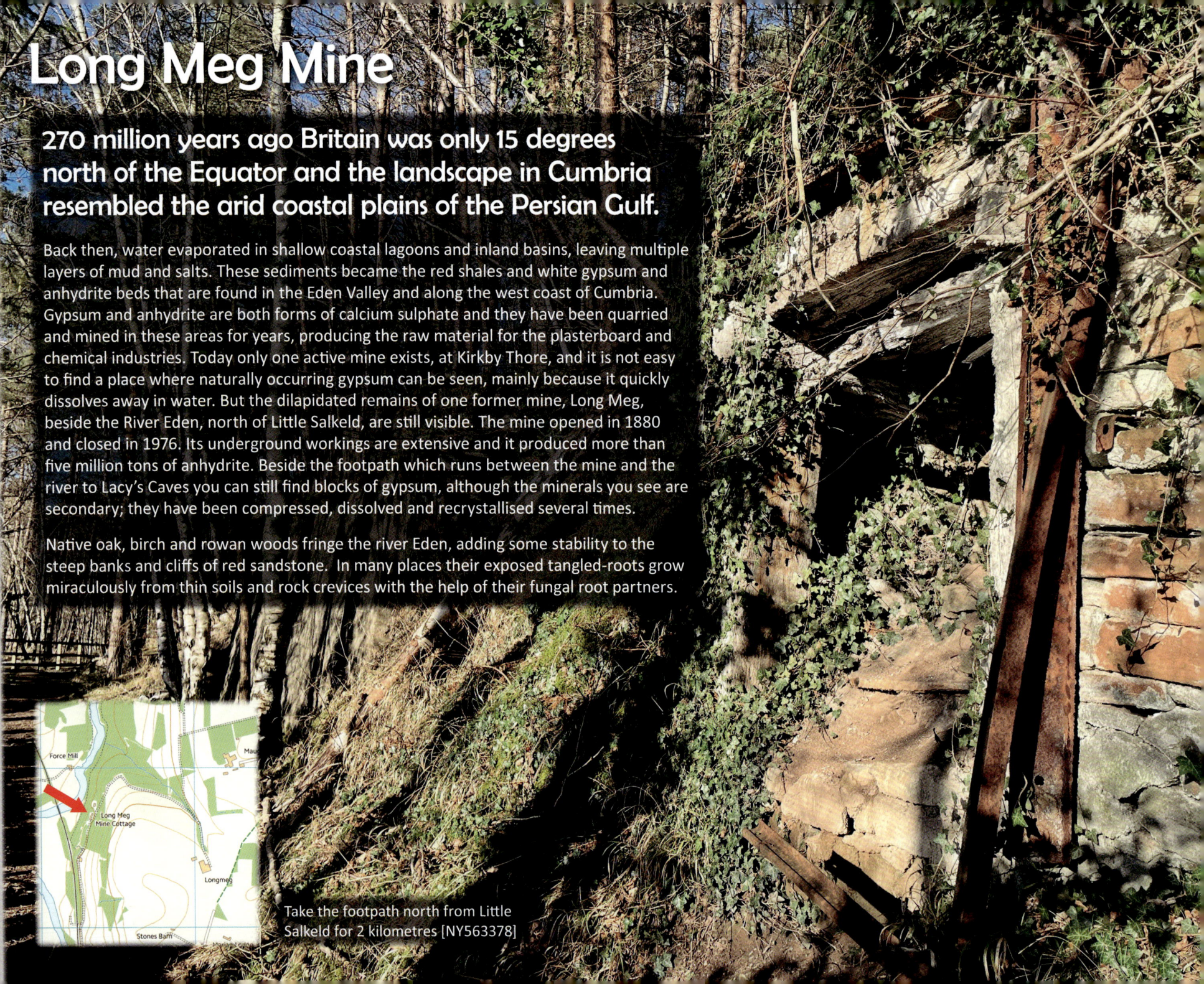

Block of gypsum and anhydrite from Long Meg mine

52

Loughrigg Fell

William Wordsworth and the poems inspired by his Cumbrian homeland are known the world over. That he influenced and was influenced by, geologists is much less well known.

One of those geologists was Adam Sedgwick, a vicar's son from Dent, (and so also a Cumbrian by today's geography!) who admitted to having no geological expertise, but was appointed as Woodwardian Professor of Geology at Cambridge University. Sedgwick spent several years in the Lake District correcting his geological deficit. His guide was Jonathan Otley, yet another Cumbrian, known to Wordsworth and an amateur geologist and clockmaker. Born in humble surroundings at Scroggs, near Loughrigg Tarn, he was a man whose knowledge of the area was encyclopaedic; he was the first to propose the basic three-fold division of Lake District rocks. While Otley remained a local hero, Sedgwick rose to be one of the foremost British geologists, who defined the Cambrian and Devonian periods of Earth's history. Sedgwick was deeply religious and that conflicted with a rapidly developing science. He once berated Darwin, a pupil of his, for the heresy of his theory of evolution.

Wordsworth and Sedgwick met and walked together many times, exchanging their respective literary and scientific views that could be landscape specific, or deeply philosophical. Much to Sedgwick's discomfort Wordsworth teased rock collectors in his poem, The Excursion. *'He who with pocket-hammer smites the edge; Of every luckless rock or prominent stone; Detaching by the stroke; A chip or splinter; To resolve his doubts; And, with that ready answer satisfied; The substance classes by some barbarous name: And thinks himself enriched; Wealthier, and doubtless wiser, than before!'* Walk over the glaciated volcanic rocks of Loughrigg Fell and you will be following in their footsteps. You can gaze across Langdale and Grasmere and admire the dramatic and ancient landscape that inspired them both.

The summit is a 2.5 kilometre walk
south and west from Rydal [NY347051]

Extract from Wordsworth on Helvellyn by Benjamin Robert Haydon. Public Domain

View west from Loughrigg Fell

53

Nenthead

Nenthead sits at 440 metres above sea level in the North Pennine hills. While it may be a quiet place now, 150 years ago it was the centre of lead and zinc mining in Cumbria.

The Victorian public buildings in this small, Quaker-built, mining town are clues to its heritage, as are the sparsely vegetated spoil heaps, the reservoirs and the mine entrances around it. Nenthead lies within the North Pennine Orefield, which was the most productive area of lead and zinc mining in Britain. Mineral veins cut across surrounding hills and valleys. Exactly why the mineral wealth occurs here is still debated. The fact that this part of the North Pennines conceals the 400-million-year-old Weardale Granite is one theory, another is linked to the magma that injected the Whin Sill 295 million years ago. The concept is that mineral-rich fluids permeated 330-million-year-old Carboniferous rocks via a rectangular pattern of faults and fractures and replaced some of the limestone layers they cut through. The ore was smelted in the valley and long stone-lined flues took the poisonous fumes away to chimneys on the hills. Other valuable minerals, like silver and cadmium, precipitated on the walls of these flues. Mining had its heyday here a century ago and eventually stopped altogether in 1961. But only ten years ago a Canadian company started prospecting for zinc. Several deep boreholes located mineral deposits and should the world price of zinc continue to rise then there may be an application to mine here again.

Meanwhile, Nenthead's mining heritage is there to experience, from the mines and smelting remains looked after and kept accessible by the Nenthead Mine Conservation Society to the many spoil heaps with their minerals and very special botany. Lead, zinc and cadmium may be toxic to most plants but some like spring sandwort, mountain pansy and Alpine penny cress have a genetic tolerance to the waste and thrive.

There is a car park 250 metres north of the mines [NY782435]

Lead and zinc ores (galena and sphalerite) from Nenthead mines.
Image by kind permission © The Trustees of the Natural History Museum, London

54

Newlands

Newlands Beck and its tributaries, Scope and Keskadale becks, flow down three of the most beautiful valleys in the Lake District.

The valleys, and the long ridges of Robinson and Hindscarth that divide them, are superb hiking country; they also have a lot to say about how the Lake District landscape evolved. In the south and east the summits of Dale Head and High Spy are volcanic ash and lavas, erupted over 450 million years ago. To the north are older Ordovician rocks, mudstones and sandstones; these were once sediments in a deep ocean separating two continents about to collide. Overprinted on this bedrock foundation are millions of years of erosion, that culminated in powerful glaciers, carving out their characteristic 'U' shaped valleys. A new technique, which uses cosmic rays to assess how long boulders moved by glaciers have been at the Earth's surface, has revealed that the last glaciers melted from these valleys less than 12,000 years ago. Since then mountain gullies and streams have transported and re-deposited debris on the valley floor.

The human influence on this landscape began with Stone, Bronze and Iron Age communities who progressively cleared the forests which had colonised the area as the climate warmed. By the 16th century man recognised another natural resource in the valleys; cutting through the mudstones were mineral veins, containing copper and lead ore but with iron and silver too. In 1564 the first Queen Elizabeth set up the 'Company of Mines Royal' to mine the ore; she needed money to balance the books and fight a war. Expert German miners were brought over and settled around Keswick. So rich was the mine that the Germans called it Gottsgabt or 'God's Gift'; we know it today as Goldscope. Quite a history for three peaceful little valleys blessed with ancient woodlands of sessile oak and carpets of moss and liverworts in their shade; and we've not even mentioned Beatrix Potter, Little Town and Mrs Tiggy-Winkle.

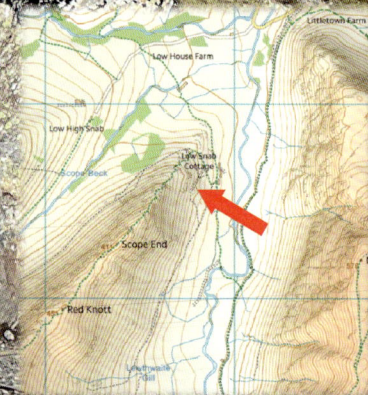

Park at Chapel Bridge south of Little Town and walk 1 kilometre through Low Snab farm to the mine [NY228186]

Looking down Little Dale from Littledale Edge

55

Roughton Gill and Silver Gill

South of Caldbeck in the fells that form the northern perimeter of the Lake District are mines for metal ores that go back centuries.

At the head of the Dale Beck valley are the steep, rocky gullies of Roughton Gill and Silver Gill. Evidence of mining is everywhere; old tunnel entrances (levels), piles of mining waste, and the remains of structures used to process the ore. These mines extracted ores that produced a large amount of lead and copper, and some silver too. Some of the workings may go back 700 years; they were certainly worked by Elizabethan miners (part of the 'Company of Mines Royal') in the 16th century and did not close until 1878. As well as the ores worked for metal over 30 different and often rare minerals have been identified here. As a result the valley is a Site of Special Scientific Interest and mineral collecting is illegal.

The geology of this area is complex. In the north are Ordovician rocks called the Eycott Volcanic Group, in the south these were cut by vertical sheets of molten magma, that produced both granite and gabbro (a rock that is usually associated with the Cuillin Mountains of Skye). Near vertical fractures in these rocks (faults) became conduits for mineral-rich fluids, which then crystallised out as veins. The veins criss-cross adjacent fells and valleys too and in Mosedale is Carrock Fell mine, one of only two places in England that produced tungsten. These hills are a quiet place now and it is hard to believe that for several hundred years they were a hive of industry, one that earned the Caldbeck Fells the accolade of being 'worth all England else'.

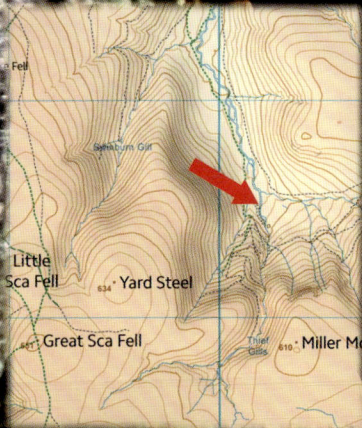

Park at Fellside and hike 3.5 kilometres south beside Dale Beck [NY302345]

Roughton Gill mines at the head of the Dale Beck valley

Saltom Pit

56

Saltom Pit shaft was sunk in 1729, it was the first mine in England to take coal from under the sea. Barely 300 metres away is Haig Colliery; it closed in 1984 and was the last pit to mine coal under the sea in Cumbria.

Saltom Pit's first engineer, Carlisle Spedding, was as creative and practical as Brunel and Stephenson. He designed a novel method of dividing the single shaft into two to ventilate the mine. He employed early atmospheric pumping engines to deal with the water that was inevitable at 130 metres beneath the sea bed. He invented a way of lighting the coal face using a flint and steel 'mill' to produce sparks which were less likely to ignite the large quantity of methane in the pit. He even had the methane piped to the surface and offered it to Whitehaven as a source of illumination. Saltom Pit closed in 1848 and only the ruins of the engine and winding house remains; recent landslides make their future uncertain.

Coal bearing rocks (Carboniferous 'Coal Measures') stretch from Whitehaven north and eastwards to Maryport and Aspatria. Over 30 seams have been mined, some reaching 4.5 metres in thickness. Coal and very large deposits of iron ore were the basis for the heavy industrial economy, and thus employment, in West Cumbria. Coal's heyday was in the last part of the 19th and the first half of the 20th centuries and in recent years it has only been extracted from a few opencast mines. Coal and other fossil fuels are recognised as the major contributors to our changing climate. Given this and a stubborn legacy of high unemployment, it is not surprising a recent application to re-open sub-sea deep mining for coal a stone's throw from Saltom Pit is the subject of heated debate.

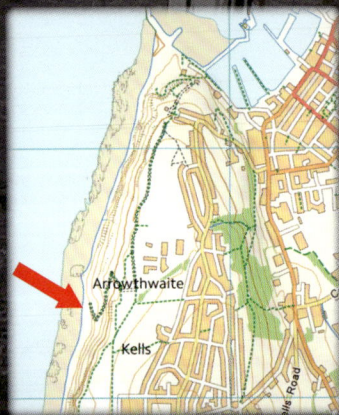

Whitehaven has rail and bus connections or you can park beside Haig Pit. The path to Saltom pit is currently closed [NY964174]

Saltom Pit and St Bees Head

57

Seathwaite

High up the western valley side are a line of waste heaps from mines that once produced one of the rarest minerals in the world: graphite.

Graphite is a form of pure carbon and because of its uses in the aerospace, nuclear, and motor industries, it is regarded as a 'critical raw material'. When the mines at Seathwaite were operating between the late 16th and mid-19th centuries, graphite was equally valued. It was first used for marking sheep, but as it behaves like Teflon its main use was for making moulds for cannon balls and coins. It also provided the raw material for the Keswick pencil factory. Such was its purity and value that the mines had armed guards to stop smuggling. The British blockade of France during the Napoleonic Wars meant the French army had to invent a substitute from graphite powder and clay.

The graphite can be in fine flakes, or in lumps. Both occur in veins and especially where the veins cross. Precisely how it formed is still debated but it probably involved high temperature, carbon-rich fluids infilling cracks in the volcanic rock which was being changed by heat and pressure (metamorphism). Graphite has been known by different names - wad, black lead and plumbago - because it was initially thought to be a form of lead. Pencils have never contained lead but we still call them lead pencils.

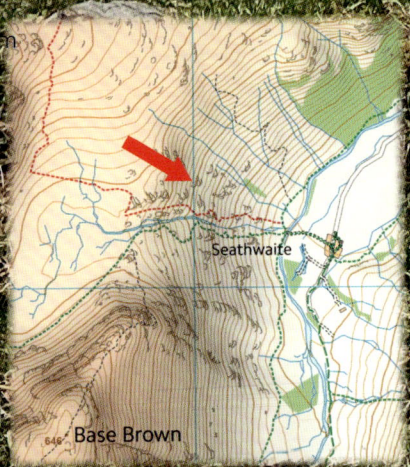

There is parking on the roadside at Seathwaite and it is a 2 kilometre steep return hike to the mines [NY232127]

John Banks Esquire stakes his claim to mine graphite in about 1752

Looking down on Seathwaite from the graphite mines

58

Sellafield

There isn't another location in Cumbria, and maybe in the UK, whose rocks have been investigated so thoroughly or argued over so much.

Between 1989 and 1997 Sellafield, more precisely Longlands Farm, was proposed by Nirex (originally the Nuclear Industry Radioactive Waste Executive) as the site of a Rock Characterisation Facility (RCF) to study the suitability of storing radioactive waste underground. Provided safety criteria can be met, geological disposal of nuclear waste was (and is) thought by many to be the optimum way to deal with a problem that is not going to go away. Sellafield is currently storing a large volume of waste at the surface.

The geology beneath Longlands Farm is essentially three layers. Up to about 100 metres of glacial sediments lie above approximately 500 metres of 250-million-year-old Triassic and Permian sandstones and mudstones. Beneath this are about 2000 metres of 450-million-year-old Ordovician volcanic rocks, a continuation of those found at the surface in the Lake District. The concept was to investigate the rocks, and the movement of water through them, in situ, to assess whether waste could be safely contained in a complex of engineered chambers within the Ordovician rocks. 65 boreholes were drilled, around 19 kilometres of drill core analysed, many geophysical surveys were done and water samples taken; multiple reports, maps and three dimensional models were made.

The plan for the RCF was rejected by the planning authority, Cumbria County Council. Nirex appealed. Following a public inquiry in 1995-96, the inspector ruled that the properties of the rocks at the site were not sufficiently well understood and rejected the appeal. The challenge of disposing of nuclear waste remains and as society grapples once more with how to resolve it, not only here but in North America and northern Europe, the geological and political issues raised more than 25 years ago are relevant again.

Take the coast path north from Seascale for 2 kilometres [NY026033]

Sellafield from Ponsonby Church

59

Tullie House, Carlisle

Tullie House Museum has outstanding Roman and Borderland exhibitions. It also has some fantastic rocks.

The Museum is home to an extensive collection of Cumbrian minerals, a legacy of the county being one of the richest mining regions in Britain. There are many other exceptional items in its collections: including 160-million-year-old Jurassic ammonites from west of Carlisle and 250-million-year-old mysterious 'vertebrate' prints left in Permian Penrith Sandstone.

It also has a large collection of Neolithic axes from the Langdale 'factory'. These are made from one very specific layer of fine-grained rock, a 450-million-year-old blue-green-grey volcanic ash called a tuff, found high up on scree slopes between Pike of Stickle and Harrison Stickle. Stone age craftsmen roughed out the axes and then honed and polished them on lower ground elsewhere; production was on an industrial scale. The finished axes have been found at archaeological sites across Britain and the unworn state of many means they are interpreted as much more than a utility item: they were ceremonial and highly valued. Their distribution also shows just how connected Stone Age society was.

I've got a confession to make. When I should have been doing my school homework in Tullie House Library Reading Room in the 1960s, I used to sneak into the museum and gaze at the birds and animals, and fossils and rocks, in their glass cases. Tullie House must take some of the credit (or blame) for the career that followed that allowed me to write this book.

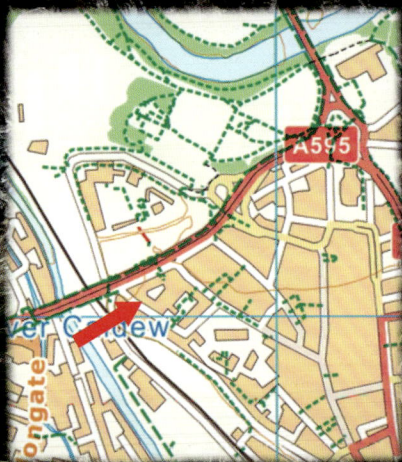

The museum is in Castle Street in the centre of the city [NY398560]

Images of Langdale axe and 'vertebrate' prints taken with kind permission Tullie House Museum, Carlisle

Fossil 'vertebrate' prints in Penrith Sandstone

60

Watchtree

We are lucky in Cumbria to have so much bedrock at the surface and in plain sight when, in the rest of England, geology is often hidden from view. But here, west of Carlisle, there is no bedrock to see, so why include this site?

Near the village of Great Orton is a new nature reserve on an old airfield. It has two concealed geological tales to tell, one recent and awful, the other about rocks that are unique in Cumbria. In 2001 Britain faced a national crisis, the foot and mouth outbreak. Millions of farm animals were destroyed. After disposal by burning failed, they were buried in mass graves; this airfield was one of these places. Understanding the geology of these pits and the way that groundwater would move was crucial if they were to be engineered so that substances leaching out did not contaminate the environment and water supplies. Monitoring of the site continues; out of sight is not out of mind.

The top layer beneath the surface of the airfield is a stony clay; glacial till (or 'boulder clay') deposited by the ice sheets 20,000 years ago. But underneath that is a rock that is only found in this small part of Cumbria. It is Jurassic and it has fossils from the time of the dinosaurs. Over 100 years ago geologists found a few scruffy bits of limestone in nearby ditches. They contained ammonites and other sea creatures (just like those in Dorset); animals which had died and been buried over 180 million years ago. In 2001 when they were excavating the foot and mouth burial pits those same fossiliferous limestones were revealed.

The potential for using the rocks beneath us (the 'geosphere') for burying things is a contentious subject. In the past our use of quarries and pits for landfill has not always been successful. But we face other challenges, like climate change and energy supply, which are linked. Both may require the use of the rocks beneath us for storage of carbon dioxide and nuclear waste. Science must underpin the evaluation but it is society which has to weigh the risks.

There is a car park beside the nature reserve visitor centre [NY310539]

The Watchtree Stone
taken from this ground

A Symbol

To the birth of
Watchtree Nature Reserve
Dedicated on this day the 7th May 2003,
the second anniversary of the final burial

A Memorial

To 448,508 sheep, 12,085 cattle, 5,719 pigs
buried here during the
Foot and Mouth outbreak of 2001

Granite erratic boulder at the entrance to Watchtree Nature Reserve

Endnote

Norman Nicholson's words on the title page hint at one thing I hope this book will do: challenge an oft-held perception that the Cumbrian landscape is largely the product of human hands, be that agriculture or architecture. Human influence on the landscape has been considerable but in the grand scheme of things, merely scratches the surface, it is cosmetic and beauty is more than skin deep.

I am, of course, biased, but it is the rocks and the processes that shaped them over almost 500 million years that are literally the bedrock of the county, the origin of its resources, the basis of its scenery and ultimately the foundation of its economy and heritage.

I hope this book helps you see Cumbria with new eyes and encourages you to explore the whole county and its diverse and beautiful places. I hope it makes you curious about what lies beneath your feet and helps you understand why these amazing places are like they are.

Grab your boots and see for yourself!

Crummock Water

The author

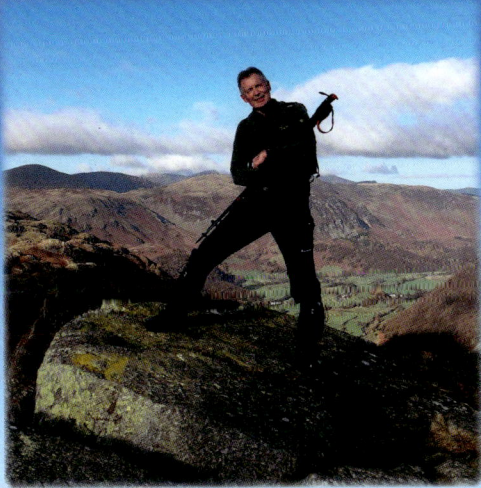

Ian Jackson was born and raised in Carlisle. His love of rocks began in the late 1950s in the Caldbeck Fells and valleys around Martindale. He has a degree in geology and geography from the University of Newcastle upon Tyne and is a Chartered Geologist and Fellow of the Geological Society.

Ian spent 18 years surveying the geology of parts of the north of England, including Cumbria, for the British Geological Survey. Later he was responsible for national and international programmes that produced the first UK, European and global digital geological maps and made them web accessible.

Ian retired from the position of BGS Operations Director in 2011 and moved to Bardon Mill in Northumberland. He hikes in Cumbria and Northumberland every week. In addition to many scientific maps, articles and reports, he is the author of Britain Beneath Our Feet, an atlas of the UK's geology, and Northumberland Rocks, a sister publication to this book.

Links to resources

Open the camera app on your phone and point at the QR code

BGS Earthwise Northern England Regional Geology

BGS Geology Viewer

BGS GeoIndex

Cumbria Wildlife Trust

Cumbria GeoConservation

Cumberland Geological Society

Westmorland Geological Society

English Lake District Geology

Geological Society of UK

Saltom Bay and Whitehaven

Cumbria
Wildlife Trust

Cumbria Wildlife Trust is a charity dedicated to safeguarding the wildlife and wild places of Cumbria - and helping people to get closer to nature. The Trust aims to bring wildlife back to the county, to help empower people to take meaningful action for nature, and to create an inclusive society where nature matters.

One of our core purposes is to help record, protect and promote a better understanding of the geological heritage of Cumbria.

The Trust manages more than 40 nature reserves across 10,000 acres. We work in partnership with local communities and others to help address the twin crises of biodiversity loss and climate change in Cumbria.

We are one of 46 Wildlife Trusts, a grassroots movement working across the UK to make life better – for wildlife, for people and for future generations. Our vision is of a thriving natural world, with our wildlife and natural habitats playing a valued role in addressing the climate and ecological emergencies, and everyone inspired to get involved in nature's recovery.

In 2022 the Trust had its 60th Anniversary. This book telling the story of Cumbria's geological heritage and its connections to wildlife is a celebration of those 60 years.